AWAKENED SCIENCE: WHERE QUANTUM PHYSICS MEETS THE DIVINE

How Modern Discoveries Validate Ancient Wisdom

By. Deilen Michelle Villegas, Ph.D.

For permissions, contact publisher:
The Shamanic Goddess, LLC
Charlotte, North Carolina, United States
www.DrDeilenMVillegas.com
Dr.Deilen.Villegas@gmail.com

Book Title: Awakened Science: Where Quantum Physics Meets The Divine

Author: Dr. Deilen Michelle Villegas, Ph.D.

Publisher: The Shamanic Goddess, LLC

Printed in the United States of America
First Edition: 2025
ISBN: 978-1-969550-04-1

Acknowledgements

To write a book is to walk with your soul in public. It is not a solitary path, but one illuminated by the wisdom, prayers, and presence of many. This book could not have been birthed without the divine orchestration of ancestors, mentors, soul family, clients, and Spirit.

To my ancestors—thank you for walking with me. Thank you for the dreams, the whispers, the rituals, and the remembrance. You held me before I could hold myself. May this work be an offering to you and to all those whose voices were silenced. I am because you are.

To my children—thank you for being my greatest teachers. Your laughter, curiosity, and resilience remind me daily why healing is sacred and why legacy matters. You are the magic in my medicine.

To my beloved husband—thank you for holding space for the sacred, for believing in my calling, and for allowing me to rise in every version of myself. Your love is an anchor in the ocean of my evolution.

To the clients, students, and community I've had the honor to walk beside—you are my inspiration. Thank you for trusting me with your stories, your healing, and your becoming. This book was written with you in mind, with your pain and power in my heart.

To my spiritual teachers and guides—seen and unseen —thank you for reminding me that I am never alone. For the downloads in the middle of the night, the divine interruptions, and the infinite confirmations.

To my soul sisters, mentors—you know who you are. Thank you for seeing me when I forgot myself, for reflecting my light when all I could see was shadow. You are mirrors of grace.

To Spirit—thank you for the gift of this life, for the codes you entrust me with, for the guidance, the mystery, and the unfolding.

And finally, to the Reader—thank you. For being willing to remember. For choosing to reclaim your power. For turning the page, not just in this book, but in your life.

May this work serve as a blessing on your journey. May you always walk in truth, in light, and in love.

With deepest reverence, Dr. Deilen Michelle Villegas

Table of Contents

About the Author

Dr. Deilen Michelle Villegas is a Board-Certified Holistic Health and Wellness Practitioner, Doctor of Philosophy in Natural Medicine, Metaphysician, and Spiritual Alchemist who bridges science, soul, and ancestral wisdom into a unified path of transformation. With over 18 years of experience in medicine—spanning Internal Medicine, Pediatrics, OB/GYN, Oncology, Triage, and Urgent Care—Dr. Villegas offers a rare fusion of clinical insight and spiritual truth.

She is the Founder and CEO of *The Shamanic Goddess, LLC*, a multidimensional wellness sanctuary and educational platform committed to mind-body-soul healing, trauma recovery, ancestral integration, and conscious leadership. Her life's work centers on helping individuals reclaim their power, rewire their minds, heal their nervous systems, and remember who they truly are at the soul level.

Raised in a spiritually blended home deeply rooted in Hoodoo, Santería, and Catholicism, Dr. Villegas was gifted early with an understanding of the unseen realms. As a Daughter of Yemayá, she carries the lineage of oceanic wisdom, emotional healing, and divine rebirth. Her near-death experience as a child marked her spiritual initiation, unlocking abilities to see, sense, and communicate across the veil.

Dr. Villegas is a published author, international speaker, trauma-informed educator, and curriculum developer. She advocates for holistic health equity and has trained thousands of practitioners, community leaders, and conscious entrepreneurs in trauma-informed care, quantum healing, energy medicine, and embodied leadership.

With credentials in Natural Medicine, Complementary & Integrative Health, CBT, NLP, Hypnotherapy, Herbalism, Clinical Sexology and Somatic Therapies, she blends ancient teachings with applied neuroscience and quantum biology to offer grounded, science-backed spiritual teachings.

Her work honors both the mystic and the modern. She believes healing is not just a personal act—it is a revolutionary one. Her mission is to help others not only survive, but thrive in alignment with their highest timeline.

Awakened Science: Where Quantum Physics Meets the Divine is both a reflection of her life's calling and a roadmap for those ready to reclaim their power, embody their purpose, and rise into multidimensional truth.

You can learn more about her offerings, courses, and community at www.theshamanicgoddess.com and www.DrDeilenMVillegas.com

INTRODUCTION:

The Sacred Reunion of Science and Spirit

There was a time when speaking about energy, vibration, or divine intelligence would brand you as a mystic—or worse, a madwoman. But today, we stand at the edge of a global awakening, where science is no longer the skeptic's shield but the mystic's ally. The discoveries emerging from the fields of quantum physics and applied neuroscience are validating what spiritual practitioners, Indigenous wisdom keepers, and metaphysicians have known all along: we are more than flesh and bone—we are energy, vibration, and consciousness in motion.

I did not learn this knowledge solely through a textbook. I came to it through lived experience—through moments of stillness where Spirit whispered through intuition, through near-death encounters that revealed hidden realms, and through years of clinical work where I witnessed the miraculous unfold in patients and clients who aligned their minds, bodies, and souls.

As a metaphysician, holistic practitioner, and someone who

walks between the worlds of ancient traditions and modern innovation, I've long understood that science and spirituality are not enemies. They are twin flames—each reflecting a different facet of the same Truth. And now, the evidence is catching up.

In this book, **Awakened Science: Where Quantum Physics Meets the Divine**, I invite you into the sacred intersection where mysticism meets measurement, and where intuition dances with intellect. This is not a book of dogma or doctrine. It is a journey—one that will challenge your perceptions, expand your awareness, and awaken your soul's memory of what you've always known but may have forgotten.

Together, we will explore:

- The quantum field as the divine matrix of reality
- The brain's role not just as a generator of thought, but as a receiver of universal intelligence
- The "clairs"—those spiritual senses often dismissed, now being studied and validated
- How ancient teachings align with neuroscience, epigenetics, and energetic medicine
- And most importantly, how you can activate your own gifts and live in alignment with both science and spirit

This book is for the seeker and the scientist within you. For the one who believes in miracles and also loves peer-reviewed studies. For those of us who understand that healing, truth, and expansion don't belong to just one tradition—they belong to all of us.

We're not here to reject science or replace spirituality. We are here to **awaken** to the knowing that both are sacred languages speaking to the same divine reality. And it's time we listen.

Welcome to the new paradigm. Welcome to Awakened Science.

CHAPTER 1: QUANTUM PHYSICS AND THE ENERGETIC UNIVERSE

"Everything we call real is made of things that cannot be regarded as real."
—Niels Bohr, Nobel Laureate in Physics

The Invisible Becomes Visible

Modern physics has shattered our once rigid understanding of the universe. The solid, material world we thought we knew—fixed, predictable, and separate—is now understood to be anything but. Beneath the surface of what we can see lies a quantum realm so extraordinary that it challenges the very nature of reality.

"The cosmos is within us. We are made of star-stuff. We are a way for the universe to know itself."

— *Carl Sagan*

At the quantum level, atoms—once believed to be the building blocks of matter—are mostly empty space. The particles that do exist don't behave like fixed objects but like **waves of energy** that fluctuate, appear, disappear, and exist in multiple places at once. We are not solid. We are vibrating, energetic beings, existing in a universe composed not of matter, but of **fields of possibility**.

This energetic reality is not theoretical. It is measurable, observable, and the foundation of technologies like quantum computing, laser optics, and nuclear medicine. But long before it was placed under a microscope, mystics, sages, and Indigenous wisdom keepers were describing this same invisible world using different language: the ether, the spirit realm, prana, chi, or Source energy. This aligns with ancient spiritual teachings that describe the universe as a dynamic, interconnected web of energy.

Superposition, Entanglement & the Observer Effect

Quantum mechanics introduced us to three revolutionary concepts that are now helping explain how spiritual phenomena might actually work:

1. Superposition:

Particles can exist in **multiple states at once** until observed. This challenges our understanding of time, space, and choice. Spiritually, this mirrors the truth that **infinite possibilities exist** until we consciously choose a path—what we might call manifestation or intention-setting.

2. Quantum Entanglement:

Particles that become linked remain connected, no matter how far apart they are—even across galaxies. A change to one particle instantly affects the other. This reflects the spiritual truth of **oneness**—that we are all interconnected through an invisible energetic web.

3. The Observer Effect:

Simply observing a quantum particle changes its behavior. Consciousness **alters matter**. What we focus on, we influence. This aligns with countless spiritual teachings: *where attention goes, energy flows.*

These aren't metaphors. They are scientific realities—backed by experiments and data. And they confirm what mystics have known: **consciousness is not a byproduct of matter—it's a force that shapes it.**

"It was not possible to formulate the laws of quantum mechanics in a fully consistent way without reference to consciousness."

— *Eugene Wigner*

Case Study: Quantum Processes in the Brain

Researchers at Trinity College Dublin have found evidence suggesting that human brain functions may involve quantum processes. By measuring brain functions related to short-term memory and conscious awareness, they detected signals indicating that these functions involve quantum entanglement. This discovery implies that consciousness could be a quantum phenomenon, bridging the gap between neuroscience and quantum physics.

The Quantum Field: The Divine Matrix

The quantum field is often described as an ocean of energy and information that surrounds and permeates all things. It contains all potentiality—what some spiritual traditions call the Akashic field or the mind of God.

Physicist David Bohm referred to it as the **"Implicate Order"**—a deeper, hidden layer of reality that gives rise to everything we see and experience. Gregg Braden calls it the **Divine Matrix** —a bridge between our inner world and the physical universe.

This concept resonates with spiritual teachings that describe a universal consciousness or divine matrix underlying all existence.

What does this mean for you? It means your thoughts, emotions, and intentions don't just float in the air. They interact with this quantum field and **send signals**—broadcasting your energy, drawing experiences, people, and opportunities into your life. You are not just living in the universe—you are **co-creating** it.

Energy Medicine, Vibration & Frequency

All matter is energy vibrating at specific frequencies. Nikola Tesla famously said, *"If you want to find the secrets of the universe, think in terms of energy, frequency, and vibration."*

In energy medicine, this principle is foundational. Sound healing, Reiki, acupuncture, and even prayer all operate on the belief that by shifting vibrational frequencies, we can heal, realign, and restore balance. Scientific studies have begun to validate these practices, showing measurable effects on the body's energy fields.

Science is now validating this with studies on:

- **Heart rate variability (HRV)** and its link to emotional and spiritual states
- **Cymatics**, showing how sound shapes matter
- **Frequency-specific microcurrent therapy**, which accelerates tissue healing

These are not fringe practices—they are **frontiers of healing**, blending ancient knowing with modern innovation.

Bridging the Past and the Future

Ancient teachings from Egypt, India, Tibet, Africa, and Indigenous cultures have long taught that the universe is alive, intelligent, and interconnected. Quantum physics is now echoing that truth through mathematical formulas and double-slit experiments.

As a metaphysician and practitioner, I see this not as a coincidence but confirmation. What was once dismissed as mysticism is now validated by physics. The veil between the spiritual and scientific has lifted.

Activation Reflection: Tuning Into the Field

> **Take a moment now. Close your eyes. Breathe deeply.**
> **Feel the vibration of your own body—your heartbeat, your breath, the subtle hum of your being.**
> **Imagine yourself as a wave, not a particle—fluid, connected, infinite.**
> **Whisper silently:** *I am part of the field. I shape the field. The field responds to me.*

When you begin to see yourself not as a separate being, but as part of the energetic universe, life transforms. You move from victim to co-creator. From seeker to source.

And this is only the beginning.

CHAPTER 2: COLLECTIVE CONSCIOUSNESS AND THE DIVINE MATRIX

"At the deepest level of reality, you and I are literally one."

—Erwin Schrödinger, Nobel Prize-winning physicist

We Are Not Alone—We Are All One

Modern science is catching up to one of the most ancient spiritual truths: **everything is connected**. Not just metaphorically, but literally—at the quantum level, we are bound by an invisible web of energy and intelligence. This invisible force field has been called many names: Source, Spirit, God, the Quantum Field, the Divine Matrix, and in science, the **Unified Field**.

In every tradition—from Indigenous cosmologies to Eastern mysticism—there is an understanding that we are all threads in the same tapestry. Now, through the study of **quantum entanglement**, **nonlocality**, and the **holographic model of the universe**, this interconnectedness is being validated by science.

"The universe is not outside of you. Look inside yourself; everything that you want, you already are."

— Rumi

We often look at the world through the lens of separation—me vs. them, self vs. other, past vs. future. But quantum science and ancient wisdom both whisper a deeper truth: **we are all connected, not just by energy, but by possibility.**

Every person you encounter is a mirror reflecting an alternate version of yourself—a possibility of who you could have been had you made a different choice, been born in a different body, under a different name, at a different time, in a different environment. They are you in another form. They represent lives you didn't live, roads you didn't take, and destinies that could still be yours.

We are not isolated observers of life—we are active participants in an **infinite unfolding of parallel potentials**. The stranger on the street, the teacher who hurt you, the child who made you smile, the ancestor who came before you—they are all part of the cosmic web of reflection and remembrance.

In this way, **we are walking through the living storybook of ourselves**, constantly encountering what our lives might look like in different forms of embodiment. That is the living expression of the **quantum multiverse**—not just across galaxies or dimensions, but here and now, **reflected in the people around us**.

Each encounter is an echo of choice:

- What would life have been if I had taken that risk?
- Who would I be if I were raised in a different culture, gender, or belief system?
- Where would I be now if I had said "yes" instead of "no," or "no" instead of "yes"?

This is how the concept of **past, present, and future** transcends time as we know it. It's not just happening on a linear scale. Time—like energy—is **nonlinear and multidimensional**. All timelines exist simultaneously, layered like pages in a book, all present here in this moment.

When we observe others without judgment, but instead with awareness, we begin to **see ourselves**. We awaken to the truth that every soul is playing a role in our spiritual evolution —sometimes to challenge us, sometimes to heal us, and sometimes to show us what's possible.

> **This is the essence of oneness:**
> Not sameness, but sacred diversity—each one of us a unique note in the cosmic symphony.
> Each life we witness is a lesson, a reflection, a revelation.
> We are not separate beings navigating a shared space.
> We are expressions of the same divine intelligence **exploring itself through infinite form.**

To live with this awareness is to walk in presence. It's to release the illusion of hierarchy or judgment and recognize the divine experiment of consciousness at play.

When you look at another, you are not just seeing them—you are seeing a **version of yourself**. And in that knowing, you unlock deeper compassion, understanding, and spiritual mastery.

The Divine Matrix: A Bridge Between Worlds

The term **Divine Matrix** was popularized by scientist and spiritual author Gregg Braden. In his work, he describes a field that connects all things, **responds to human emotion**, and serves as a bridge between the spiritual and the material worlds.

"There is a field, beyond all notions of right and wrong. I'll meet you there."

— Rumi

This concept is deeply rooted in quantum field theory, which describes a sea of fluctuating energy that underlies all matter. The **Higgs Field**, for instance, is what gives subatomic particles their mass. But it's not static—it responds to energy inputs. Our intentions, emotions, and consciousness become **vibrational signals** that interact with this field. This is not pseudoscience

—this is cutting-edge physics and neuroscience converging on something spiritual communities have always known: **we co-create our reality.**

The **Divine Matrix** is not a myth, nor merely poetic language —it is the fabric of existence, the field that connects all consciousness and energy, all matter and spirit. It is the space between worlds—the invisible realm through which intention, emotion, and frequency interact with physical reality.

In quantum physics, this field is known in various forms: the **Zero-Point Field**, the **Unified Field**, the **quantum vacuum**. To the spiritually attuned, it is the place where prayer travels, where intuition is born, and where miracles are made. It is both **cosmic infrastructure** and **spiritual intelligence**—the very canvas upon which the universe paints itself into being.

> **Gregg Braden** describes the Divine Matrix as the "**container that holds the universe, the bridge between all things, and the mirror that shows us what we have created.**" In other words, it is the realm that **receives our energy and reflects it back to us**, shaped by the quality of our consciousness.

Every thought, feeling, and belief you emit is like a frequency-coded message sent into this matrix. And just like sound waves resonate with matching vibrations, the matrix mirrors back realities that are in alignment with your internal state. This is **not philosophy—it is physics**. Studies in quantum mechanics and field theory are showing that **energy is responsive**, and more importantly, **consciousness is causative**.

Activation Reflection: Connecting with the Divine Matrix

> **Find a quiet moment.**
> **Breathe slowly and deeply.**
> **Visualize yourself surrounded by a glowing network of light, stretching infinitely in all**

directions.
See every thought, every prayer, every heartbeat pulsing through this grid—connected to all beings, seen and unseen.
Whisper: "I am never alone. I am part of the Whole. My intention is a ripple in the quantum sea."

The Spiritual Science of Resonance

When we say "like attracts like," we are speaking to the principle of **resonance.** Just as two tuning forks vibrate in harmony when brought into proximity, so too do frequencies of thought and feeling draw matching experiences through the matrix.

- When you embody love, you begin to resonate with loving opportunities.
- When you harbor fear or scarcity, the field reflects situations that affirm those vibrations.
- When you raise your consciousness, you shift your **energetic GPS**—drawing in higher outcomes and synchronicities that feel miraculous but are actually mathematical.

This is why practices like **prayer, affirmation, visualization, and meditation** work. They **tune your personal frequency** into harmony with the outcomes you wish to create. You're not manipulating the universe—you're **aligning with it**.

Cross-Cultural Echoes of the Divine Matrix

Every spiritual tradition has a name for this field, this bridge between the visible and invisible:

- In **Kemetian (Ancient Egyptian)** cosmology, it is the **Nu (Nun)**—the primordial waters of chaos and creation, from which the gods and life itself emerge.

- In **Yoruba tradition**, it is **Ashe**—the divine life force that lives in all things and carries intention into manifestation.

- In **Vedic texts**, it is **Akasha**—the subtle energy that holds all memory, vibration, and truth.

- In **Christian mysticism**, it is the **Holy Spirit**, the breath of the Divine that intercedes and moves throughout creation.

- In **modern physics**, it is the **quantum field**, described as a non-local, energetic substrate of infinite potential.

Different names. Same truth. These are not metaphors—they are **maps** of the same territory, interpreted through the lens of culture and time.

We Are the Bridge

Here is the most powerful truth of all: **you are not separate from the Divine Matrix. You are the Matrix.** Your body, your mind, your spirit—they are not just moving through this field—they are made of it. The bridge between worlds lives within you.

When we pray, we're not reaching outward—we're reaching **inward and upward** into the very field that birthed us. When we heal, we're not imposing a change—we're **repatterning our vibration**, allowing the Divine Matrix to mirror wholeness back to us.

This is why miracles are possible. Not because they defy natural law, but because they **activate higher laws**—laws of energy, resonance, and conscious co-creation. And every one of us is encoded with the ability to access this field.

Activation Reflection: Becoming the Bridge

> **Place your hands over your heart. Close your eyes.**
> **Breathe.**
> **Imagine a radiant golden light connecting your heart to the stars above and the Earth below.**
> **Say aloud or in your spirit:**
> *I am the bridge between the visible and invisible.*
> *I am the sender and receiver of divine signals.*
> *The field responds to my frequency.*
> *I align with truth, love, and sacred power.*
> And so it is.

Case Study: The Global Consciousness Project

At Princeton University, researchers launched a bold initiative called the **Global Consciousness Project (GCP).** Using a network of random number generators (RNGs) located around the world, they tested whether **mass human emotion** could influence physical systems.

They discovered that during **major global events**—such as 9/11, natural disasters, and moments of worldwide meditation or prayer—the RNGs displayed **statistical anomalies**, suggesting that **collective emotional energy** influences the field of probability itself.

> **"There seems to be a global consciousness, an intelligence that we all share that has measurable effects on the physical world."**
> —Dr. Roger Nelson, Director of the GCP

This study suggests that our **shared emotional and mental states** have the power to influence the very structure of reality —a concept that both ancient shamans and modern quantum theorists would recognize.

Consciousness Is Not Local—It's Universal

Traditional neuroscience long held that consciousness was

produced by the brain. But more and more, we are seeing evidence that consciousness may actually **exist beyond the brain**, and that the brain functions as a **receiver**, not a generator.

This theory is supported by the works of:

- **Dr. Karl Pribram**, who developed the **holographic brain theory**, suggesting memory and perception are nonlocal and field-based
- **Sir Roger Penrose** and **Dr. Stuart Hameroff**, whose **Orch-OR model** posits that **quantum processes** in the brain may allow access to universal consciousness

These theories align beautifully with spiritual ideas of the **Higher Self**, the **Collective Unconscious** (Carl Jung), and **spiritual downloads**—the moments when we receive insights, visions, and messages from beyond.

"You are not a drop in the ocean. You are the entire ocean in a drop."

—Rumi

Modern science has begun to dismantle the long-held belief that consciousness originates in the brain. Instead, we're seeing that the brain may function more like a **receiver**—a bioelectric instrument that tunes into frequencies of thought, inspiration, and divine intelligence that already exist in the **quantum field**.

This idea is not new to mystics, metaphysicians, or shamans. They have long known that **consciousness exists beyond the body**, and that the mind can access information far beyond time, space, or logic. This is why people experience prophetic dreams, intuitive downloads, ancestral visitations, or even memories that cannot be explained through linear time. Science is catching up with this truth through studies in **non-local consciousness**, **telepathy**, **remote viewing**, and **shared spiritual phenomena**.

But here's where things go deeper—and darker.

Power, Control, and the Weaponization of Frequency

Those in positions of power—governments, corporations, elite institutions—have long understood the **laws of energy, vibration, and frequency**. They know how these forces shape reality. They understand that **words carry energy**, that **music alters brainwave states**, and that **emotion is the magnetic force of manifestation**.

And rather than use this knowledge to **liberate**, they have often used it to **manipulate**.

Through mass media, entertainment, and public messaging, they influence the **energetic frequency of the collective**:

- **Music** tuned to unnatural frequencies like **440 Hz**, rather than the heart-centered 432 Hz, has been shown to create agitation and dissonance in the body.

- **Movies** are strategically released with violent, dystopian, or fear-based narratives that trigger the brain's survival centers and reinforce scarcity, division, and disempowerment.

- **News cycles** are designed to keep people in a state of anxiety, rage, and helplessness—frequencies that make individuals more susceptible to control and less connected to their divine intuition.

This is not a conspiracy—it is **energetic warfare**. And it works because **consciousness is not confined to one mind—it is collective**. By controlling the vibration of the masses, they maintain control over the reality we all co-create.

The Age of Aquarius and the Fall of the Old Paradigm

But something extraordinary is happening.

We are entering the **Age of Aquarius**—an astrological era associated with awakening, enlightenment, truth, and innovation. This is the age of **collective consciousness**, **divine rebellion**, and the **rise of the inner healer, the mystic, the artist, and the truth-teller**.

As the vibration of the planet increases, the old tactics of fear and control begin to fall away. People are **waking up**, questioning everything, and no longer willing to live inside systems of greed, oppression, and division.

More and more **artists, actors, musicians, and creators** are awakening to their roles as spiritual activists. They are using their platforms, their music, their scripts, and their voices to speak truth, to pierce the veil, to awaken hearts. This threatens the systems of control that once dominated the narrative—and the energy.

We're seeing this reflected in music that honors ancestral roots, in films that unveil spiritual truths hidden in myth, and in viral messages of healing, unity, and empowerment that are spreading like wildfire.

The Role of TikTok and Digital Downloads

Ironically, one of the greatest catalysts for this collective awakening came through something that was **initially designed to distract**: the app **TikTok**.

Originally engineered for quick dopamine hits through dances and soundbites, TikTok has **evolved into a digital portal for truth, spiritual downloads, and collective connection**. People from across the world, across languages and belief systems, began to **receive the same messages**—messages about frequencies, about soul contracts, about healing ancestral trauma, about the Great Awakening.

We've witnessed thousands of individuals from separate countries and cultures, who've never spoken to one another, share nearly identical prophetic dreams, intuitive insights, and spontaneous revelations. This isn't a coincidence—it's **coherence**. It's what happens when a collective consciousness begins to sync, vibrate, and **remember who it is.**

> **We are remembering that we are divine.**
> **We are remembering that we are creators.**
> **We are remembering that we are one.**

TikTok became more than an app—it became a **mirror**. It showed us that our desires for love, truth, peace, healing, and justice are **not unique—they are universal**. It reminded us that we are far more **similar than we are different**, and that many of our belief systems, while called by different names, are rooted in **the same sacred longing**: to be whole, to be seen, and to live in harmony.

The Time of Awakening Is Now

As more individuals raise their frequency, the veil continues to thin. The illusions—of separation, of race, of religion, of division —begin to dissolve. And what is revealed is the truth:

We were never meant to be enemies. We were never meant to be asleep. We were always meant to rise.

This moment is not a glitch. It is not chaos without purpose. It is the sacred unraveling. And you—reading these words—are part of that unraveling. You are not just here to witness it. **You are here to anchor it.**

Activation Reflection: Reclaiming Your Frequency

> **Close your eyes. Breathe in through the nose, out through the mouth.**
> **Place your hands over your solar plexus, your power**

center.
Say aloud or within:
I reclaim my energy.
I command my frequency.
I align with truth.
I vibrate love, not fear.
I receive divine insight.
I am no longer controlled by illusion.
I am sovereign. I am free. I am light.
And so it is.

Akashic Records and Universal Memory

In many spiritual traditions, there is a belief in a field that records all thought, action, and emotion—a divine database known as the **Akashic Records**. The concept is strikingly similar to what some physicists call the **Zero-Point Field**—a background energy field with virtually infinite information and potential.

> **"The vacuum is not empty. It's a plenum of all possible realities."**
> —Dr. Ervin Laszlo, systems theorist and founder of the Akashic Field Theory

If consciousness and memory exist in this field rather than solely in the brain, then when we meditate, pray, or engage in spiritual practices, we are **tapping into this greater intelligence** —remembering rather than learning, aligning rather than seeking.

> *"The Akashic Records are the soul's journey across time—a divine archive of every thought, action, and experience."*
> *— Dolores Cannon*

In spiritual and metaphysical traditions, the **Akashic Records** are often described as a vast, ethereal library that holds the energetic imprint of every soul's journey. Accessing these records allows individuals to reconnect with their **Higher Self**, unveiling the truths of who they are, who they have been, and the potential of who they can become.

Reconnecting with the Soul's Blueprint

Throughout our lives, experiences such as trauma, societal conditioning, and personal challenges can cause us to drift away from our authentic selves. Engaging with the Akashic Records —through practices like hypnotherapy, guided meditation, or deep introspection—facilitates a **soul awakening**, helping us remember our original path and purpose.

This process is not about discovering something new but about **remembering**. It's a journey back to the essence of our being, shedding layers of forgetfulness to embrace the truth of our existence.

Children and the Thin Veil of Memory

Many spiritual beliefs suggest that children, especially in their early years, remain closely connected to the spiritual realm. Their innocence and lack of extensive life conditioning allow them to access memories and perceptions beyond the physical world.

One interpretation posits that the **white tunnel** often reported in near-death experiences is the same light encountered during birth—a transition point between realms. The initial cries of a newborn might signify the soul's mourning for its previous existence, a poignant farewell to a past life.

Scientific Investigations into Past-Life Memories

The phenomenon of children recalling past lives has been studied extensively by researchers like Dr. Ian Stevenson and Dr. Jim B. Tucker at the University of Virginia. Their work has documented numerous cases where children provided detailed accounts of previous lives, which were later verified through historical records.

In about 35% of these cases, children had birthmarks or physical anomalies corresponding to wounds or injuries from the individuals they claimed to have been in past lives

Case Studies Highlighting Past-Life Memories

- **James Leininger**: An American child who, from a young age, recounted detailed memories of being a World War II fighter pilot. His descriptions of aircraft, naval ships, and events were later corroborated with historical facts.

- **Shanti Devi**: A girl from India who, at the age of four, began speaking of a past life in a town she had never visited. She provided names, addresses, and details that were later confirmed by the family of the deceased individual she claimed to have been.

These cases, among many others, suggest that the soul retains memories across lifetimes, accessible under certain conditions or through specific practices.

The Collective Awakening and Shared Memories

In our interconnected age, platforms like TikTok have become unexpected conduits for spiritual awakening. Individuals from diverse backgrounds share experiences of past-life memories, spiritual insights, and intuitive downloads, often finding resonance with others worldwide.

This collective sharing reinforces the idea of a **universal consciousness**, where souls are interconnected, and memories or insights are not confined by geography or culture.

Embracing the Journey of Remembrance

Accessing the Akashic Records and embracing past-life memories is a journey of self-discovery and healing. It allows individuals to understand recurring patterns, heal ancestral wounds, and align with their soul's purpose.

> **Activation Reflection**:
> **Find a quiet space. Close your eyes and take deep, calming breaths. Visualize a grand library filled with infinite knowledge.** *As you walk through, feel drawn to a particular book—your soul's record. Open it and allow the memories, lessons, and truths to flow into your consciousness. Embrace the insights with love and gratitude.*

What This Means For You

You are not isolated. Your thoughts are not random. Your prayers are not whispers into a void. You are part of a **collective consciousness**—a living, breathing intelligence that responds to you and moves through you.

This is why intention-setting, group meditation, ancestral rituals, and collective prayer are so powerful. When done together, they don't just make us feel good—they **amplify frequencies**, shift timelines, and change the energetic fabric of reality.

"The wound is the place where the Light enters you."

— Rumi

Understanding that consciousness is universal—and that we are divine extensions of that consciousness—completely changes the way we live, think, and heal. It means that you are never separate from truth, from guidance, or from Source. But to access that truth, you must be willing to do something that society has trained you to avoid: **go within**.

In a world that thrives on noise, distraction, and external validation, **inner work is revolutionary**. Introspection is often dismissed or pathologized, but in truth, it is the highest form of self-liberation. You've likely been taught to look outward for answers: to the media, to systems, to people in power, to labels, even to social media likes. But your greatest compass—your most accurate GPS—lives inside of you.

This is why practices like **reflection, journaling, shadow work, somatic healing, and meditation** are not just wellness trends —they are ancient technologies that help you **reconnect to the memory of who you truly are**.

You are not broken.
You are not behind.
You are not lost.
You are simply **remembering**.

The System Was Never Built For You to Find Yourself

The painful truth is that much of society is intentionally designed to keep you disconnected from that inner wisdom. The systems we navigate—education, healthcare, government, religion—are often structured in ways that **discourage critical thinking**, **shame self-reflection**, and **punish emotional depth**.

You're told:

- To suppress your intuition in favor of "facts"
- To seek validation from authorities who have never lived your truth
- To measure your worth by how well you fit into boxes that were never made for you

But how can you become your highest self using systems that were never created with your wholeness in mind?

This is why reclaiming your power starts with **radical self-inquiry**. Not just asking, *Who am I?* but *Who told me who I was*

supposed to be?
Not just wondering *What do I want?* but *Why did I learn to desire this?*
And not just seeking purpose, but **remembering that your existence is the purpose.**

Your Soul Has Always Known

The guidance you're looking for? You already have it.
The answers you seek? You've dreamed them, sensed them, ignored them, and are now being called to receive them.
Your soul remembers who you were before the world told you to forget.

When you sit in stillness, when you dare to face your own reflection, when you feel the stirrings of your truth rise up—**that is the Akashic current flowing through you.** That is your divine inheritance being activated.

You don't need permission to heal.
You don't need credentials to be worthy.
You don't need a system to approve of your awakening.

All you need is the courage to come home to yourself.

The Path Forward

Doing this work is not easy. You will confront wounds, unravel beliefs, and break free from narratives that have kept you safe but small. But in doing so, you'll become **sovereign**—and that is the most dangerous and beautiful thing you can be in a system designed for conformity.

When you reclaim your frequency, you reclaim your life.
When you remember your divinity, you remember your purpose.
When you do the inner work, you rise—not just for yourself, but for generations before and after you.

You are not just healing you.

You are healing your bloodline, your community, your timeline, and the collective frequency of the planet.

Activation Reflection: Returning to the Temple Within

Find a quiet place. Sit or lie down with your spine supported.
Close your eyes. Breathe deeply into your belly.
Whisper within:
I remember who I am.
I release the stories that were never mine.
I reclaim my truth.
I return to the temple within.
I am safe. I am sacred. I am seen.
And so it is.

CHAPTER 3: APPLIED NEUROSCIENCE AND THE SPIRITUAL BRAIN

"The mind is not a vessel to be filled, but a fire to be kindled."
— Plutarch

The Brain is Not Just a Processor—It's a Portal

In our modern world, the brain is often described like a machine—a biological computer that stores data, calculates information, and controls behavior. But what if the brain is more than just hardware? What if it's a **conduit**, a receiver, a translator for something much greater?

Spiritual traditions have long taught that **insight, intuition, and divine inspiration** do not originate in the mind—they pass *through* it. In this view, the brain is not the source of consciousness—it is the **receiver**. Neuroscience is beginning to confirm this, revealing that states of deep meditation, spiritual activation, and even emotional breakthroughs involve parts of the brain **receiving and integrating information** in highly complex, non-linear ways.

This is the sacred intersection of applied neuroscience and spirituality—where ancient practices are now being **backed by brain scans, neurofeedback, and peer-reviewed research**.

Brainwaves: The Frequencies of Awakening

The human brain operates on multiple frequency states known as **brainwaves**, and each one corresponds to a unique level of consciousness:

- **Beta (12-30 Hz):** Active, alert thinking—associated with logic, problem-solving, and stress
- **Alpha (8-12 Hz):** Relaxed, meditative, calm—often accessed during mindfulness and quiet reflection
- **Theta (4-8 Hz):** Deep meditation, intuition, dream state—where the subconscious and superconscious meet
- **Delta (0.5-4 Hz):** Deep sleep and healing—where the body regenerates and spirit travels
- **Gamma (30-100 Hz):** Heightened cognition, bliss states, unity consciousness

Practices like **breathwork, meditation, chanting, visualization, and energy healing** help shift the brain from high-frequency Beta into more spiritually receptive states like Theta and Gamma. These are the frequencies where people report receiving **downloads**, seeing visions, experiencing telepathy, and connecting with ancestors or higher realms.

> *"Meditation is not about becoming blank—it's about tuning your brain into the frequencies of divine insight."*

The Pineal Gland: The Third Eye in Science

Often referred to in spiritual traditions as the **Third Eye**, the **pineal gland** is a small endocrine gland located deep in the center of the brain. It regulates melatonin and circadian rhythms, but spiritually, it's been revered as the **seat of the soul**.

Modern research shows the pineal gland contains **piezoelectric crystals**, which means it may respond to pressure and frequency. It's also rich in **DMT** (Dimethyltryptamine), a naturally occurring compound in the human body known to induce mystical and interdimensional experiences.

Activating the pineal gland through meditation, breathwork, or energy work can lead to profound **altered states of consciousness**. Many believe this is the gateway to the Akashic field, to remembering our past lives, and to perceiving non-physical realms.

> *"The pineal gland is the antenna—when you raise your frequency, you strengthen the signal."*

Neuroplasticity and the Power of Healing the Mind

> *"You can't heal a body you hate, or rewire a mind you ignore. Healing begins with remembrance—and remembrance requires presence."*

One of the most profound discoveries in modern neuroscience is the principle of **neuroplasticity**—the brain's remarkable ability to change, adapt, and rewire itself throughout life. This completely revolutionized our understanding of healing. It means that your thoughts are not just passive experiences—they are **active creators of your internal world**.

For those who have endured trauma, chronic stress, or negative conditioning, this is a sacred truth: **you are not doomed by your past—you are rewiring toward your future.** And with consistency, intention, and compassion, you can quite literally **reshape your mind, reframe your life, and reclaim your wholeness.**

Let's explore three powerful modalities that harness the sacred science of neuroplasticity: **Cognitive Behavioral Therapy (CBT), Mindfulness, and Positive Affirmations.**

Cognitive Behavioral Therapy (CBT): Rewiring Thought Loops

CBT is a science-backed, structured approach that centers

on identifying and interrupting destructive thought patterns. Think of it as spiritual hygiene for the mind—it empowers you to stop letting unconscious beliefs dictate your behavior and begin consciously choosing thoughts that align with your truth.

Key Components:

- **Identification of Negative Thoughts**: Often our internal narratives are not just negative, but inherited—from family, society, trauma, or survival. CBT helps you shine a light on these shadow beliefs.

- **Cognitive Restructuring**: Once seen, these thoughts can be challenged and reframed. For example, "I always fail" becomes, "I have learned from every setback, and I am growing stronger."

- **Behavioral Activation**: Change isn't just mental—it's embodied. CBT encourages taking aligned action, creating experiences that reinforce the new belief systems you're cultivating.

> *"You don't think your way into a new way of living. You live your way into a new way of thinking."*

Mindfulness: Presence as a Portal for Healing

Mindfulness is more than a buzzword—it is an ancient, evidence-based practice that directly affects **brain structure and chemistry**. It teaches you to become the **observer of your thoughts** rather than the prisoner of them.

Benefits for the Brain:

- **Awareness and Acceptance**: Through mindful witnessing, you learn to see your thoughts and emotions clearly, without attachment or shame.

- **Stress Reduction**: Mindfulness reduces cortisol levels and calms the amygdala, the part of the brain associated with fear and emotional reactivity.

- **Focus and Emotional Regulation**: Studies have shown that consistent mindfulness practice increases gray matter density in areas of the brain responsible for decision-making and empathy.

Mindfulness rewires your brain to return home to the present—where your power, peace, and possibilities live.

Positive Affirmations: Speaking Life Into the Mind

Every word you speak is a frequency. Every thought you repeat becomes a belief. **Positive affirmations are not wishful thinking—they are spiritual commands broadcasted into the quantum field and reinforced through the neural network.**

Neuroscience of Affirmations:

- **Self-Affirmation Theory** supports that positive statements about the self activate the reward centers in the brain and **counteract threat-based thinking**.

- **Neural Rewiring** occurs when affirmations are repeated with emotion and visualization, forging new synaptic connections that support confidence, resilience, and joy.

- **Emotional Resilience** grows stronger as your inner dialogue shifts from shame to self-compassion, from lack to abundance.

"Every affirmation is a permission slip to remember your divinity."

The Sacred Process of Mindset Reframing and Reprogramming

Healing the brain is a sacred practice—and it's one you can begin anytime. Here's a simplified flow for **reframing your mindset and reprogramming your consciousness** using the principles of neuroplasticity:

1. Awareness

Begin with observation. Notice the repeated thought loops or emotional triggers. Ask yourself: *Whose voice is this? Is this belief mine—or inherited?*

2. Challenge

Interrupt the autopilot. Question the belief. *Is it true? Has it always been true? What's the cost of holding onto this thought?*

3. Reframe

Create a new belief that feels supportive, honest, and empowering. *I'm not behind—I'm on divine time. I'm not broken—I'm becoming.*

4. Practice

Write it. Speak it. Visualize it. Practice the new thought daily through affirmations, meditation, journaling, and action.

5. Integrate

Live it. Reinforce your new belief with aligned behaviors. When the old patterns arise, don't judge them—reaffirm the new ones.

Activation Reflection: Rewire in Real Time

> **Close your eyes. Place your hand on your heart. Breathe deeply.**
> **Bring to mind a limiting belief that has been**

holding you back.
See it clearly—don't resist it, just witness it.
Now speak aloud, or in your spirit:
I see this story.
I honor its roots.
But I choose a new truth.
I am safe to grow. I am worthy of joy. I am free to rise.
And so it is.

Spiritual Experiences in the Lab

Groundbreaking research by Dr. Andrew Newberg (neurotheologian and author of *How God Changes Your Brain*) has used fMRI technology to study the brains of people engaged in deep prayer, meditation, and spiritual ritual. His findings show:

- Increased activity in the **frontal lobes**—linked to focus, compassion, and problem-solving
- Decreased activity in the **parietal lobes**—associated with the sense of separateness and ego
- Elevated production of **dopamine and serotonin**, fostering joy, peace, and connection

These experiences mirror those described by mystics, monks, shamans, and seekers—proving that spiritual practice isn't just emotional or energetic—it's **neurological**.

The Body Remembers, But It Also Rewrites

Applied neuroscience also confirms that **somatic memory**—emotional trauma stored in the body—can be released through intentional movement, breath, sound, and sensory work. Modalities like **EMDR, EFT (tapping), yoga nidra, and trauma-informed breathwork** all leverage the brain-body connection to rewrite internal narratives.

This is where **spirituality and neuroscience merge most powerfully**: you can not only access past imprints through altered states—you can also **reprogram the mind** to align with your soul's true frequency.

Activation Reflection: Awakening the Mind of Spirit

> **Place both hands on your temples. Breathe deeply and slowly.**
> **Visualize a stream of golden light flowing down from the crown of your head into your brain.**
> **Imagine that light illuminating every corner —cleansing trauma, activating wisdom, igniting insight.**
>
> **Whisper:**
> *I am attuned to divine frequencies.*
> *My mind is a sacred vessel.*
> *I awaken my gifts with love and intention.*
> *My brain is wired for joy, peace, and spiritual power.*
> And so it is.

Visualizing the Spiritual Brain

Case Study 1: Carmelite Nuns and Mystical Union

In a groundbreaking study, neuroscientist Dr. Mario Beauregard and his team at the University of Montreal used functional magnetic resonance imaging (fMRI) to examine the brains of Carmelite nuns as they relived profound mystical experiences. The research revealed that these spiritual recollections activated multiple brain regions, including areas associated with emotion, self-awareness, and body perception. This suggests that mystical experiences are complex and involve a network of brain regions rather than a single "God spot".

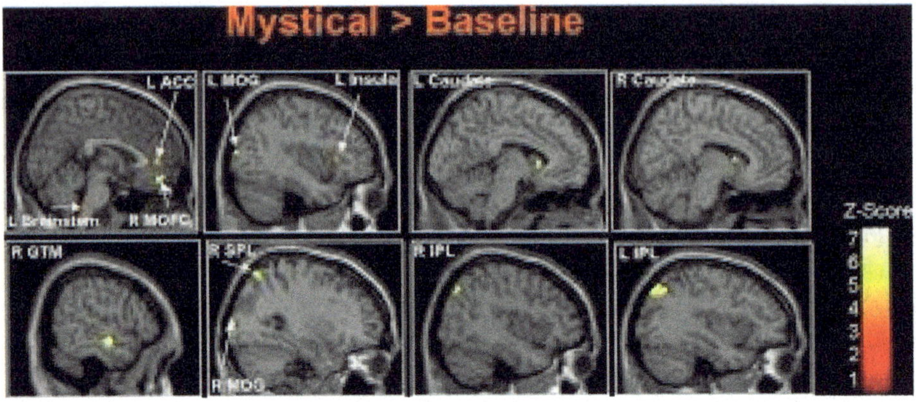

Case Study 2: Devout Mormons and the Brain's Reward System

A study led by neuroscientist Dr. Jeffrey Anderson at the University of Utah explored how spiritual experiences affect the brain's reward circuitry. Using fMRI scans, the research team observed that when devout Mormons engaged in prayer and religious activities, there was significant activation in the nucleus accumbens—a region associated with feelings of pleasure and reward. This activation pattern is similar to the brain's response to music, love, and even drug use, indicating that spiritual experiences can produce profound feelings of joy and reinforcement.

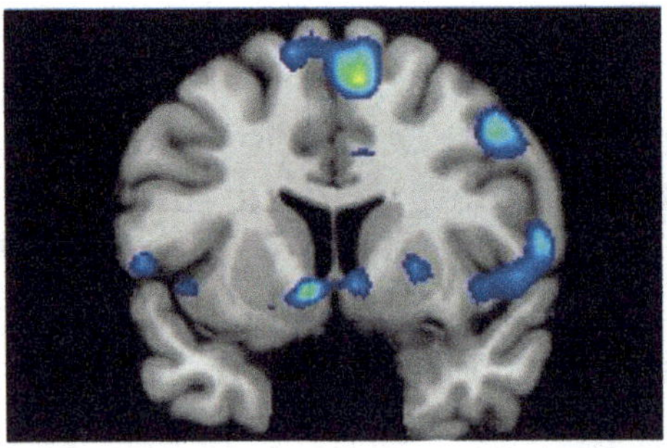

Case Study 3: Orgasmic Meditation and Altered States

Dr. Andrew Newberg, a prominent figure in neurotheology, conducted a study on practitioners of orgasmic meditation

(OM), a practice combining mindfulness and intimate touch. Using neuroimaging techniques, Newberg found that OM induced brain activity patterns similar to those observed during traditional meditation and religious experiences. Specifically, there was altered activity in the frontal and temporal lobes, regions associated with emotional processing and self-awareness. Participants reported feelings of connection, warmth, and relaxation, suggesting that OM can facilitate states of consciousness akin to spiritual experiences.

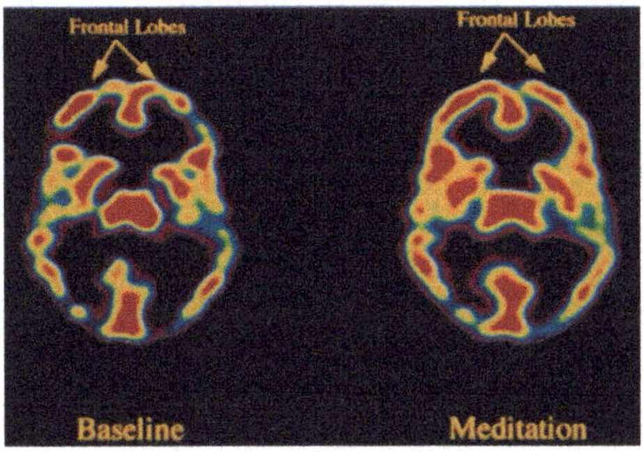

These visual representations provide compelling evidence of the tangible neural correlates of spiritual experiences.

Activation Reflection: Harmonizing Science and Spirit

> **Close your eyes and take a deep breath. Visualize your brain illuminated with light, each region harmoniously engaged in a symphony of spiritual awareness. Feel the connection between your thoughts, emotions, and higher consciousness. Affirm:**
> *I am attuned to the divine frequencies within me.*
> *My brain is a vessel for spiritual insight and transformation.*
> *I embrace the unity of science and spirit in my journey toward enlightenment.*
> And so it is.

Implications for Personal Transformation

> *"Healing doesn't begin with medicine. It begins with remembering that you are energy, and energy can transform."*

Understanding the profound influence of thoughts, emotions, and spiritual awareness on the brain opens the door to a radically empowering truth: **you are not simply a body that thinks—you are a field of consciousness that creates.** And when you begin to heal and rewire the brain with intention, you don't just improve your mood or mental health—you initiate a **soul-level transformation** that ripples through every cell of your being.

This new paradigm of healing is being embraced through the emerging field of **quantum medicine**—a revolutionary approach that bridges neuroscience, energetic anatomy, and spiritual awareness into one coherent, holistic system.

Quantum Medicine: A Healing Paradigm for the Future

Quantum medicine is not science fiction—it's **science evolved**. Rooted in the principles of quantum physics, it acknowledges that the human being is not just a collection of tissues and systems, but a **vibrational being surrounded and influenced by a dynamic energy field** known as the **biofield**.

This biofield is sensitive to:

- **Thoughts and beliefs**
- **Emotions and stress levels**
- **Environmental frequencies**
- **Spiritual alignment or misalignment**

When our thoughts are disempowered, our emotions

unprocessed, or our energy is out of sync with our higher truth, it creates incoherence in the biofield—eventually manifesting as dis-ease in the physical body.

Quantum medicine views illness not as a singular event, but as a **message from the field**, asking us to look deeper into the energetic and emotional roots of imbalance. It is the intersection where **modern science meets ancient wisdom**.

Core Principles of Quantum Healing

1. You Are Energy First, Matter Second

Every part of your being—from cells to organs—is vibrating with energy. Healing must occur **not only on the physical level**, but in the energetic realm where illness often begins.

2. Consciousness Affects Reality

Your thoughts, emotions, and intentions influence not just your mindset but your biology. This is the power of **epigenetics**—showing that genes are not destiny, but respond to the **internal environment** of the mind and the external energy field of your life.

3. Illness Is Information

Symptoms are signals. Pain is not the enemy—it is a communicator. When we begin to see illness as an expression of energetic disharmony, we gain the power to **heal at the root**, not just treat at the surface.

Quantum Tools and Spiritual Technology

Quantum medicine doesn't rely on pharmaceuticals alone. Instead, it integrates technology and metaphysics to create **multi-dimensional healing experiences**. These include:

Biofeedback Devices

Tools that measure your body's subtle frequencies, giving

insight into energetic stressors, imbalances, and blockages in real time.

Frequency Generators

Used to emit healing frequencies that re-harmonize dissonant energy patterns in the body, much like tuning a musical instrument.

Guided Visualization & Hypnotherapy

Practices that alter brainwave states and access subconscious imprints for energetic rewiring and emotional release.

Energy Healing Modalities

Practices like **Reiki**, **Qigong**, **ThetaHealing**, and **Shamanic Energy Medicine** work directly with the biofield, clearing stagnation and restoring alignment with one's higher self.

These aren't simply "alternative"—they are **restorative**. They work with the body's natural intelligence rather than imposing control upon it.

A Truly Holistic Model

Quantum medicine mirrors the ancient holistic model of healing: mind, body, and spirit are inseparable. Your inner world creates your outer condition. Your vibration attracts your circumstances. And your healing is a **journey of remembrance,** not just recovery.

This approach emphasizes:

- **Personalized care**: No two people carry the same energetic blueprint. Treatments are tailored to the **unique spiritual, emotional, and biological matrix** of the individual.

- **Preventive awareness**: Energy field imbalances can often be sensed and addressed long before physical symptoms arise, allowing for **preventive healing**.

- **Empowered participation**: The patient is not passive —they are a co-creator in their healing journey. This restores **sovereignty** to the individual.

This Is the Revolution of Healing

We are not waiting on the future of medicine—we are becoming it.
When you combine neuroscience with energy medicine, quantum physics with spiritual intelligence, you create a healing paradigm that is **limitless**, **intuitive**, and deeply **human**.

Your brain is not the limit. Your energy is not bound.
Your healing is not just recovery—it's an **activation** of your most divine self.

Activation Reflection: Becoming a Quantum Healer of Your Own Life

> **Place both hands over your heart and breathe into the center of your chest.**
> **Visualize your body surrounded by a golden-white field of light.**
> **Feel that light scanning, clearing, and elevating every part of you—mind, body, and soul.**
> **Whisper aloud or within:**
> *I am energy in motion.*
> *I am consciousness expressed through form.*
> *I heal with awareness. I align with truth.*
> *My body listens. My spirit leads. My mind transforms.*
> *I am my own medicine.*
> And so it is.

Pioneers of Quantum Medicine

The field of quantum medicine has been shaped by several visionary thinkers who have bridged the gap between science and spirituality:

- **Dr. Bruce Lipton**: A cellular biologist renowned for his work in epigenetics, Dr. Lipton's research emphasizes how beliefs and emotions can influence gene expression, suggesting that our perceptions can directly affect our biology.

- **Dr. Joe Dispenza**: Combining neuroscience, quantum physics, and epigenetics, Dr. Dispenza teaches how meditation and mental rehearsal can rewire the brain, leading to profound personal transformations and healing.

- **Lynne McTaggart**: An investigative journalist and author, McTaggart's work explores the interconnectedness of all things through the lens of quantum physics, emphasizing the power of intention and collective consciousness in healing.

- **Dr. Amit Goswami**: A theoretical physicist who advocates for a paradigm shift in science, Dr. Goswami integrates consciousness into the framework of quantum physics, proposing that consciousness is the foundation of all existence.

- **Dr. Paul Drouin**: Founder of Quantum University, Dr. Drouin has been instrumental in developing educational programs that merge traditional medicine with quantum physics, promoting a holistic approach to healthcare.

Comparative Overview: Traditional Medicine vs. Quantum Medicine

This comparative chart illustrates the fundamental differences between traditional and quantum medicine, highlighting the latter's emphasis on the interconnectedness of all aspects of human existence and the role of consciousness in health and healing.

Aspect	Traditional Medicine	Quantum Medicine
Philosophy	Focuses on treating specific symptoms and diseases.	Emphasizes holistic healing of mind, body, and spirit.
Approach	Reactive; intervenes after disease onset.	Proactive; aims to prevent illness by maintaining balance.
Diagnosis	Based on physical symptoms and tests.	Considers energy fields and consciousness levels.
Treatment Modalities	Pharmaceuticals, surgery, and radiation.	Energy healing, meditation, and frequency therapies.
Patient Role	Passive recipient of care.	Active participant in the healing process.
View of the Body	Sees the body as a machine with separate parts.	Views the body as an interconnected energy system.
Integration of Mind	Often separates mental health from physical health.	Integrates mental and emotional states into overall health.
Scientific Basis	Grounded in classical physics and biology.	Incorporates principles of quantum physics and consciousness.
Outcome Measurement	Relies on measurable physical changes.	Includes subjective experiences and energy shifts.

CHAPTER 4: CHAKRA ENERGY ANATOMY – THE BODY'S VIBRATIONAL MAP OF HEALING

"Your body is not just physical—it is a sacred architecture of light, vibration, and memory."

Chakras as the Energy Network of Consciousness

The chakra system is more than an esoteric symbol. It is a **vibrational map of consciousness**, encoded into the architecture of the human body. Rooted in ancient Indian yogic philosophy and validated by modern science through subtle energy research, the chakras represent the energetic centers that govern our physical health, emotional balance, spiritual awareness, and behavioral patterns.

In both mystical and clinical frameworks, these seven primary energy centers correlate with **nerve plexuses, endocrine glands, and electromagnetic activity**. When energy flows freely through these points, we are aligned with our highest potential. When blocked or distorted, we experience dis-ease—physically, emotionally, or spiritually.

Let us now explore each chakra as a **spiritual gate and scientific system**—a synergy of vibration and biology.

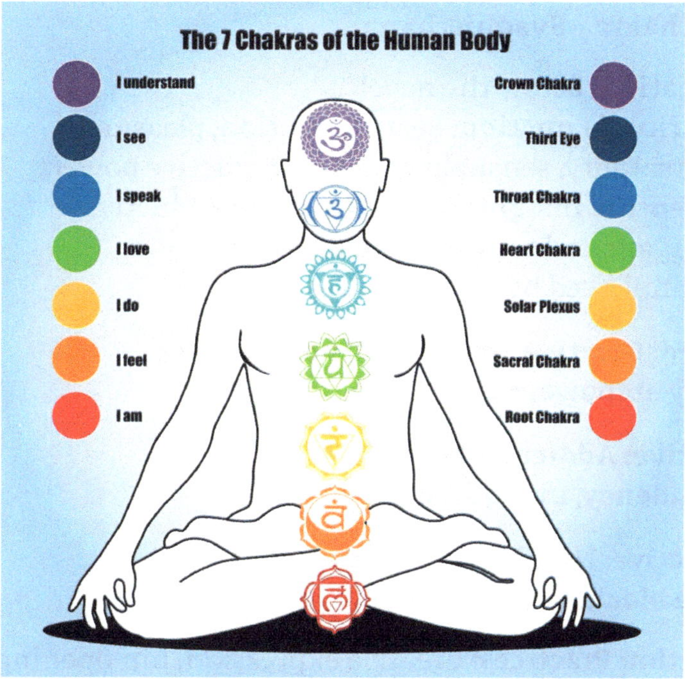

Root Chakra – Muladhara

- **Location:** Base of spine
- **Spiritual Function:** Anchors the soul into physical reality; our relationship with survival, tribe, and the Earth.
- **Scientific Insight:** Governs the adrenal glands and pelvic nerve plexus; influences cortisol levels and fight-or-flight response.

Balanced: Grounded, safe, financially secure, stable.

Overactive: Controlling, rigid, materialistic, anxious.

Underactive: Disconnected, fearful, insecure, dissociative.

Restoration Practices: Grounding meditation, barefoot walking, root vegetables, pelvic-focused movement.

A regulated root creates a sanctuary for the soul to safely reside in the body.

Sacral Chakra – Svadhisthana

- **Location:** Below the navel
- **Spiritual Function:** Seat of emotion, pleasure, creativity, sensuality, and reproductive power.
- **Scientific Insight:** Linked to the reproductive organs, kidneys, and sacral nerve plexus; influenced by estrogen/testosterone.

Balanced: Creative, emotionally expressive, sexually empowered.

Overactive: Addictive tendencies, codependency, mood swings.

Underactive: Numbness, repression, creative block, lack of libido.

Restoration Practices: Creative expression, hip-opening yoga, sacred sexuality rituals, water therapy.

> *The sacral chakra reminds us that pleasure is not a sin—it is a sacred signal of vitality.*

Solar Plexus Chakra – Manipura

- **Location:** Upper abdomen
- **Spiritual Function:** Personal willpower, confidence, motivation, and purpose.
- **Scientific Insight:** Regulates digestion via the pancreas and enteric nervous system ("gut brain"); closely tied to the vagus nerve.

Balanced: Empowered, confident, disciplined, clear.

Overactive: Aggressive, perfectionist, controlling.

Underactive: Indecisive, self-doubting, powerless.

Restoration Practices: Sunlight exposure, breath

of fire, intention-setting, core activation.

> *When the solar plexus is aligned, your inner fire fuels your soul's mission.*

Heart Chakra – Anahata

- **Location:** Center of chest
- **Spiritual Function:** Love, compassion, forgiveness, unity consciousness.
- **Scientific Insight:** Corresponds with the thymus gland and heart plexus; HeartMath research confirms heart's electromagnetic field influences coherence, emotion, and immunity.

Balanced: Open-hearted, empathetic, loving.

Overactive: Overgiving, martyrdom, emotional burnout.

Underactive: Guarded, bitter, emotionally detached.

Restoration Practices: Gratitude rituals, heart-centered meditation, forgiveness journaling, breathwork.

> *An open heart is not weakness—it is the strongest frequency of healing.*

Throat Chakra – Vishuddha

- **Location:** Throat
- **Spiritual Function:** Truth, self-expression, communication, sacred speech.
- **Scientific Insight:** Linked to the thyroid gland and vocal cords; governs hormonal balance and respiratory rhythm.

Balanced: Clear communicator, authentic, confident in expression.

Overactive: Talkative, interruptive, domineering.

Underactive: Silent, timid, withholding truth.

Restoration Practices: Vocal toning, journaling, chanting, honest conversations.

> *Your truth is your medicine. When voiced, it heals you and liberates others.*

Third Eye Chakra – Ajna

- **Location:** Between the eyebrows
- **Spiritual Function:** Intuition, perception, imagination, foresight.
- **Scientific Insight:** Associated with the pineal and pituitary glands; sensitive to circadian rhythms, melatonin, and light perception.

Balanced: Intuitive, insightful, visionary.

Overactive: Overanalyzing, delusional, fantasy-prone.

Underactive: Lack of direction, poor memory, spiritual disconnection.

Restoration Practices: Meditation, dreamwork, journaling insights, visualization.

> *Your third eye is not a mystical concept—it is your inner lens to the quantum field.*

Crown Chakra – Sahasrara

- **Location:** Top of the head
- **Spiritual Function:** Divine connection, spiritual awareness, enlightenment.
- **Scientific Insight:** Influences brainwave states and potential interaction with non-local consciousness via quantum entanglement theories.

Balanced: Peaceful, inspired, spiritually connected.

Overactive: Spiritual bypassing, disembodiment.

Underactive: Cynical, disconnected, existential dread.

Restoration Practices: Silence, prayer, contemplation, time in nature, devotional practices.

> *The crown is not about escaping the world—it's about seeing it through the eyes of Source.*

Science Meets Spirit: Validating the Chakra System

- **Dr. Hiroshi Motoyama** found electrical changes in chakra areas using biofeedback.
- **Kirlian Photography** and **Gas Discharge Visualization (GDV)** capture biofields influenced by chakra-focused meditation.
- **HeartMath Institute** affirms the measurable energetic coherence of the heart field.
- **Korotkov's research** showed biofield restoration after Reiki and chakra-based interventions.

Embodying Balance: The Chakras in Daily Life

Chakra healing is not just metaphysical—it's practical. To embody balance:

- **Live in integrity.** Untruths block the throat. Resentments constrict the heart.
- **Listen to your body.** Chronic gut issues may point to solar plexus imbalance. Tight hips may store sacral trauma.
- **Nurture all parts of you.** Your physical, emotional, mental, and spiritual layers are sacred.

When your chakras are aligned, you are not just balanced—you are **fully embodied** as your divine human self.

You remember that you are energy. That your body is a blueprint of spirit. That your healing is multidimensional.

And in that remembrance... you return to power. you return to presence. you return to Source.

And so it is.

Reflective Questions: How Do Your Chakras Speak to You?

Use the following prompts to tune in to your energy system. Let your responses guide your healing and integration:

1. **Root Chakra:** When do I feel the most grounded or the most unsteady? What brings me a sense of physical safety?
2. **Sacral Chakra:** How do I express or suppress pleasure and creativity? What is my relationship with sensuality?
3. **Solar Plexus Chakra:** Do I trust myself to take action? Where do I give away my power or over-control others?
4. **Heart Chakra:** Where am I withholding love—from myself or others? What would forgiveness feel like in my body?
5. **Throat Chakra:** Do I speak my truth freely? What truths am I afraid to express?
6. **Third Eye Chakra:** How often do I honor my intuition? What blocks me from trusting my inner knowing?
7. **Crown Chakra:** What practices connect me to something greater? Do I allow space for silence, awe, and presence?

As you reflect, listen gently. The chakras don't demand—they whisper. When you respond, they open.

CHAPTER 5: THE BRAIN AS A RECEIVER, NOT JUST A GENERATOR

"The brain is the hardware. Consciousness is the signal. And Spirit is the source."

Beyond the Brain: Where Thought Really Begins

For decades, mainstream science insisted that consciousness was a product of brain activity—that neural firing created our awareness, emotions, and sense of self. But mounting evidence suggests the opposite: **the brain does not create consciousness —it channels it**.

Think of your brain as a **radio**. It doesn't produce the music —it **tunes into it**. The frequency you tune into determines what you hear. In the same way, your brain receives, decodes, and transmits **energetic signals**—thoughts, emotions, intuitive insights—from a larger field of intelligence.

This field is what many refer to as the **quantum field**, **Source energy**, the **Akashic Field**, or simply **Universal Consciousness**. And just as a finely tuned instrument makes better music, a balanced, awakened brain becomes a **clearer channel** for higher frequencies of thought, wisdom, and guidance.

Scientific Theories Supporting the Receiver Model

Several researchers, physicians, and theorists support this non-local model of consciousness, including:

- **Dr. Karl Pribram**, with his *Holographic Brain Theory*, proposed that the brain operates more like a lens or decoder for a greater holographic reality, processing energy and information that exists outside of our limited perception.

- **Dr. Rupert Sheldrake** introduced the concept of **Morphic Resonance**, suggesting that memory and behavior are shaped by collective fields of information—not stored in the brain, but accessed through it.

- **Sir Roger Penrose** and **Dr. Stuart Hameroff** developed the **Orch-OR Theory**, proposing that quantum processes occur in the brain's microtubules, allowing consciousness to be more than biochemical—it's **quantum mechanical**.

These ideas challenge the materialist view of the brain and suggest what mystics, shamans, and sages have always known: **the mind is more than the body, and the soul speaks in frequencies beyond five senses.**

Intuition: The Language of Universal Intelligence

> *"Your intuition is your higher self whispering through the body. It speaks in nudges, not noise. In knowing, not logic."*

Your intuition isn't random. It isn't silly. It isn't made up.

It is your soul's **sacred compass**, guiding you through the physical world with divine insight and energetic precision.

Our body is not just flesh and bone—it is a **sacred vessel** for spirit. A multidimensional suit designed to carry consciousness through the physical realm. And within this vessel lives a highly sensitive, intelligent guidance system: **intuition**.

When you feel that internal nudge, that whisper in your mind, that flutter in your gut, that moment of knowing without knowing how—you're not imagining it. You are **receiving**.

Your intuition is how your **Higher Self**, your **spirit guides**, and the **collective field of divine consciousness** communicate with you. It bypasses logic and reason. It doesn't argue or explain. It simply **knows**.

How Intuition Shows Up in the Body and Mind

Intuition manifests in various forms depending on your energetic wiring:

- **A sudden urge** to call someone, change direction, or walk away
- **A tightness in your chest or stomach** when something feels "off"
- **A light, excited sensation** when you're aligned with something meant for you
- **A recurring thought, vision, or idea** that gently persists
- **A chill or tingle** that confirms spiritual truth or presence

These signals aren't coincidences. They are communications from the field—**downloads** delivered in the language your body understands.

Why Society Teaches Us to Distrust It

From a young age, we are taught to **prioritize intellect over instinct**, to follow rules instead of intuition, and to silence our inner voice in favor of external authorities. This conditioning disconnects us from our innate spiritual wisdom. Why?

Because a society rooted in control, conformity, and capitalism cannot thrive when its people are **sovereign, intuitive, and spiritually aware**.

We're bombarded with distractions—news cycles, entertainment, overstimulation, social conditioning—all designed to dull our spiritual senses. We're taught to ignore the gut feeling that says "this isn't right," or the inner knowing that something more awaits us.

But here's the truth: **your intuition is your original technology**. And it is always waiting for you to turn the volume back up.

Strengthening the Intuitive Signal

Like a muscle, your intuition becomes stronger the more you use it. The more you listen to it, the more accurate, specific, and aligned it becomes. And the more you **ignore** it, the more the world feels chaotic, unclear, or out of sync.

Developing intuition is not about escaping fear—it's about **learning to differentiate fear from divine guidance.**

- **Fear-based doubt** feels heavy, anxious, and filled with inner conflict.
- **Intuitive nudges** feel neutral, clear, and firm—even when they don't "make sense" right away.

Learning to **validate your emotions**, understand your energy, and discern your internal voice is essential to spiritual embodiment. This is not just about "trusting your gut"—it's about **reclaiming the ancient language of your soul**.

A Return to Spiritual Sovereignty

As we remember how to receive, we begin to see that intuition was never meant to be an afterthought—it was always meant to be the first voice we heard.

Honoring your intuition is a **radical act of self-trust**. It is a return to your divine self. It is the highest form of personal liberation in a world that profits off your doubt.

When you follow your intuition:

- You move in alignment with your soul's timeline
- You attract opportunities that match your energetic truth
- You avoid misaligned situations that drain your power
- You begin to **live from your center—not the chaos of the world**

Activation Reflection: Trusting the Voice Within

> **Close your eyes. Place one hand on your heart, one on your solar plexus.**
> **Breathe deeply. Tune in. Recall a moment you followed your intuition—how did it feel?**
> **Now say aloud or within:**
> *I trust myself.*
> *I honor the voice within.*
> *I am divinely guided, always.*
> *My body is wise. My soul is clear.*
> *I walk the path of inner knowing.*
> And so it is.

Dreams, Downloads, and the Akashic Cloud

> *"Your soul doesn't learn—it remembers."*

Every dream, vision, intuitive message, and sudden inner knowing you experience is a form of **communication with the field**—the divine intelligence that surrounds and connects all beings across time, space, and dimensions. Like a cloud storing universal memory, this energetic archive is known in metaphysical circles as the **Akashic Records**.

Much like your phone syncs with the cloud to download stored data, your spirit does the same through the subconscious mind, the dream state, meditation, and altered consciousness. These moments of *knowing without learning*, of *seeing without having seen*, of *hearing the whisper of truth in your own voice*—these are **downloads**.

The Akashic Field: The Soul's Internet

The **Akashic Field** is a concept found in many spiritual traditions—an invisible library of all human and soul experiences across lifetimes and timelines. Every thought, emotion, decision, and karmic imprint exists in this energetic matrix. We access it in moments of:

- Deep sleep and REM dreaming
- Meditation or trance
- Shamanic journeying
- Past-life regressions
- Sudden insights or intuitive hits
- Moments of divine synchronicity

This is why you sometimes "just know" what to do, or why you meet someone and feel like you've known them for lifetimes. These are moments where your **spirit is syncing** with the cloud of soul memory—your personal energetic archive.

Deja Vu: A Glitch—or a Divine Signal?

One of the most mystifying and common experiences of accessing non-local memory is **Deja Vu**.

That strange, disorienting moment where time seems to fold in on itself—where you feel like you've been here before, said this before, lived this exact moment already. The conversation, the scenery, the sensation—it feels **too familiar**, too perfect, too exact.

"Wait... haven't I done this before?"

In mystic and quantum communities, **Deja Vu** is not a malfunction of the brain—it is a **signal from the soul**. A sign that your spirit has **shifted timelines**, merged realities, or crossed a multidimensional bridge.

Deja Vu and Timeline Shifting

In the language of quantum theory, there are infinite potential timelines—each representing a different version of your life based on the choices you've made, and those you have yet to make.

Deja Vu may occur when:

- You make a powerful choice that **redirects your trajectory**, placing you on a timeline you've "previewed" before
- You **collapse a quantum potential** into reality—a moment your soul knew would eventually arrive
- You shift from one energetic state of being into another, **merging multiple versions of yourself**
- Your soul receives a **download from a parallel self** that has already lived through this moment

In other words, Deja Vu is not a glitch—it's a **gateway**. A soft reminder that reality is not linear. That you are not just walking forward in time, but also shifting **across** time.

Deja Vu as Confirmation, Not Confusion

Rather than questioning your sanity or brushing it off, you can begin to view Deja Vu as:

- A **confirmation** that you're on a path your soul recognizes
- A **signal** that a quantum choice or activation is unfolding
- A **downloaded memory** from a parallel self or alternate version of you
- A moment of **energetic alignment** with your highest self or karmic timeline

What we call "weird" is often just **divine intelligence expressing itself through the veil**.

Dreams and Parallel Realities

Much like Deja Vu, our dreams are often portals to other timelines—windows into our subconscious and multidimensional selves. Within dreamtime:

- You may **process karma** from this life or another
- You might **travel to other realms** to receive spiritual instruction
- You may even **encounter other versions of yourself**, living different lives in parallel realities

When we sleep, we release the conscious mind's grip on time, logic, and physical form. In this liminal state, we enter the Akashic Cloud—receiving healing, messages, preparation, and guidance.

Activation Reflection: Decoding the Signals of the Soul

Close your eyes. Visualize yourself surrounded by streams of light, like fiber optic cords connecting you to infinite timelines.
See yourself choosing alignment with your highest path.

Whisper aloud or within:
I honor the sacred signals.
I trust the knowing that comes from within and beyond.
My soul remembers.
My spirit leads.
I shift timelines with intention.
I walk the path already etched in the stars.
And so it is.

Collective Consciousness and Shared Insight

> *"When the student is ready, the teacher appears. When the world is ready, the ancestors rise."*

The idea of **collective consciousness** suggests that humanity shares a unified field of intelligence—a vibrational blueprint where thought, emotion, intention, and memory are not isolated to the individual, but echoed across the collective. And while this concept has lived in quantum theory and metaphysics for decades, we are now witnessing it unfold in real time across the globe.

In these last few years, **something has shifted**.

We are seeing a mass awakening—a spiritual migration back to **ancestral truth**, **indigenous wisdom**, and **divine remembrance**. People of all ethnicities, religions, and nationalities are feeling the pull—not to something new, but to something *ancient* within them.

A cellular knowing. A soul recall.

A call from the ancestors: **"Come home."**

A Global Soul Awakening

Since the years between **2019 and 2021**, particularly during the quiet chaos of the COVID-19 pandemic, something profound occurred. With the world shut down and our distractions stripped away, millions experienced a **spiritual implosion**—what mystics call the **Dark Night of the Soul**.

We were forced inward.

Many felt as if they were losing their minds. Old identities began to die. Religious doctrines no longer resonated. Systems we trusted began to crumble. The trauma we buried began to surface. And without realizing it, we were being **prepared**—not broken.

> It was not madness.
> It was initiation.

Those who awakened during this time were not chosen because they were better—they were simply **called first**, because they were meant to be the **way-showers**, the **midwives of a new earth**. The *guides*, *practitioners*, *healers*, and *messengers* who would later hold space for those just beginning their own awakening.

This wave of awakening was not isolated—it happened in **massive synchronicity**:

- People began having visions, dreams, and downloads—often with the same themes and symbols

- Ancestors began appearing in meditations and dreams, calling their descendants to remember

- Collective rituals and spiritual circles began forming across continents

- People began seeking out herbal medicine, astrology, ancestral prayer, divination, and energy healing—even if they'd never been exposed to it before

This is the embodiment of collective consciousness in motion. **One frequency, pulsing through millions of souls at once.**

The Call to Return to Ancestral Wisdom

In nearly every spiritual lineage, there is a prophecy that speaks of this time—a time when **the children would return to the old ways**. A time when the ancestors would awaken within us, guiding us to remember:

- The medicine of the earth
- The songs of our lineage
- The rituals of our people
- The truth of who we are beyond colonization, religion, and cultural amnesia

Even for those who don't know their bloodline or whose history has been disrupted by erasure, colonization, or displacement—the call still arrives. Because ancestral wisdom is not only passed through DNA—it is passed through **spirit**.

> *You don't have to know their names to remember their essence.*
> *You don't have to know the language to hear their voice.*

From Ridicule to Reverence

What's remarkable about this moment is that those who once felt *misunderstood, mocked, or marginalized* for their spiritual beliefs are now becoming **the very teachers, space-holders, and lightbearers** the collective is turning toward.

Once labeled as "too sensitive," "too much," or "crazy," these individuals are now:

- Holding sacred space as trauma-informed guides
- Creating new systems of holistic healing
- Reclaiming indigenous and diasporic wisdom
- Teaching others how to journey through awakening without fear

> We have gone from **seeking safety in silence** to becoming the **embodied wisdom our ancestors prayed for**.

The world is beginning to see that the mystics were never mad—they were ahead of their time.
And now, *time has caught up*.

Why Now? A Mystery in Divine Timing

There may not be a singular explanation for why this is happening now, and so fast. But in spiritual and quantum traditions, **frequency determines manifestation**—and right now, the collective frequency is rising.

As more people awaken, the collective field shifts, making it easier for others to awaken as well. The veil thins. The guidance becomes clearer. The synchronicities multiply.

And while the world may seem chaotic, the truth is: **a new world is being born.**

Activation Reflection: Answering the Call

Place your hands over your heart and breathe.
Visualize your ancestors standing behind you—dozens, hundreds, thousands of souls.
Feel them pouring wisdom, light, and protection into your body.

Now say aloud or within:
I remember who I am.
I honor those who came before me.
I answer the call.
I walk the path of truth, healing, and remembrance.
I am the prayer and the fulfillment.
I am the voice and the vessel.
I am the return.
And so it is.

Activating the Receiver: How to Tune Into Higher Consciousness

To become a clearer channel for divine intelligence, we must quiet the noise, strengthen the signal, and trust the message.

Tools to Activate the Receiver Brain:

- **Meditation**: Creates stillness to hear and feel intuitive insight

- **Visualization**: Trains the brain to perceive alternate realities

- **Breathwork**: Alters brainwaves and expands perception

- **Sound Healing**: Vibrationally retunes the body and field

- **Journaling & Automatic Writing**: Captures intuitive downloads from the subconscious

- **Sacred Plant Medicine (if aligned and safe)**: Expands consciousness and accesses deep soul wisdom

When practiced consistently, these tools **fine-tune your frequency**, making it easier to receive messages from your higher self, spirit guides, ancestors, or Source itself.

You Are More Than You've Been Told

You are not your thoughts—you are the observer of thought. You are not your brain—you are the consciousness using it to navigate this dimension.

And when you begin to live as a receiver, not just a thinker, you open yourself to a vast, divine intelligence that already knows the path, the purpose, and the truth of who you are.

Activation Reflection: Tuning In

Sit quietly. Breathe deeply.
Visualize your brain surrounded by a radiant field of light, like a cosmic satellite.

Whisper:
I open myself to divine insight.
I trust the messages I receive.
I am a vessel of wisdom.
I am a receiver of truth.
My mind is open. My heart is aligned. My spirit is listening.
And so it is.

Prophecies of the Return: Ancestral Wisdom Across Cultures

Throughout time, Indigenous elders, spiritual leaders, and mystic traditions have shared visions of a future where humanity would be called back to its origins—not to regress, but to **remember**. Across continents, there are recurring themes:

- A collapse of old systems
- A rise in spiritual consciousness
- A return to the Earth
- And the **awakening of the ancestors** through their descendants

These were not just myths or metaphors—they were **spiritual roadmaps**, encoded in ritual, story, and blood memory.

The Eagle and the Condor Prophecy – Andean & North American Indigenous

This sacred prophecy, shared by Indigenous peoples from the Amazon and Andes, speaks of a time when humanity would split into two paths:

- The **Condor** representing the **heart, intuition, and Indigenous wisdom**
- The **Eagle** representing the **mind, technology, and Western knowledge**

For 500 years, the Eagle was said to dominate, creating imbalance and disconnection from nature. But the prophecy foretells a time—the **21st century**—when the Eagle and the Condor would fly together once again, symbolizing a **reunion of ancient wisdom and modern insight**, and a return to balance.

> This is the time of prophecy. The time of the return.
> And the flight has already begun.

The Seventh Fire Prophecy – Anishinaabe People (Turtle Island)

The Anishinaabe people of North America carry the **Seven Fires Prophecy**, a roadmap through seven epochs of time. The **Seventh Fire**, the age we are currently in, speaks of a moment when:

- A new people will rise
- They will retrace the steps of their ancestors
- They will reclaim the sacred ways
- And the **Elders' teachings will return to the hearts of the young**

> "If the new people remain strong in their quest, the Water Drum of the Midewiwin Lodge will again sound its voice. There will be a rebirth of the Anishinaabe Nation and a rekindling of old flames."

This prophecy directly echoes the current movement: **descendants returning to their roots**, choosing healing over forgetting, and **carrying forward the original medicine of their people.**

Zulu Prophecy – South Africa

The Zulu nation, through oral traditions preserved by **Credo Mutwa**, a Zulu Sangoma and High Sanusi (spiritual elder), speaks of a **"Star Seed Awakening"**, where children of Africa and the diaspora would begin to awaken across the globe:

- Remembering the wisdom of the ancestors
- Reclaiming their connection to the stars
- Uniting their fragmented identity across oceans and borders

This awakening was said to **begin slowly, then rise rapidly**—and would coincide with global unrest, pandemics, and natural upheavals.

> "The ancient drums will call again. The blood of Africa will rise in the hearts of her scattered children."

The Hopi Prophecy – North America (Southwest)

The Hopi elders share a prophecy of the **"Great Purification"**, a time when:

- The Earth would be out of balance
- Humanity would face a choice between sacred life or destruction
- Those who remembered the Earth, the spirits, and the ancestors would rise to restore harmony

The prophecy describes the return of the **"True White Brother,"** a metaphor for people of all colors returning to sacred wisdom—those who live by heart, spirit, and the sacred web of life.

> "When the sacred hoop is mended, and the four races unite with wisdom, we will once again live in harmony with the Earth."

Maori Prophecy – Aotearoa (New Zealand)

The Maori speak of **"Ka hinga atu he tētēkura, ka hara mai he tētēkura"**
(*When one fern frond dies, another arises to take its place*)—a metaphor for ancestral knowledge never being lost, only passed on.

Their prophecy points to a **time of global realignment**, where their moko (descendants) would rise up with the knowledge of the atua (gods/ancestors) to:

- Heal the land
- Reclaim sacred traditions
- And **merge ancient knowing with future pathways**

> "We are our ancestors, returned in new skin, carrying old fire."

Dogon People – Mali, West Africa

The Dogon have long preserved ancient cosmological knowledge, passed down through secret oral traditions. Their prophecy speaks of a time when humanity will return to its **original vibration**, and the Nommo—spirit ancestors—will guide humanity back to harmony with **Sirius energy** (spiritual consciousness).

This aligns with the **rising interest in Kemetic spirituality, astrology, rootwork, and ancestral healing** throughout the African diaspora today.

> "You will know it is time when the spirit of Sirius calls back the bloodline of the stars."

The Second Coming of Christ – Christian Prophecy Reinterpreted

For centuries, the Christian Bible has spoken of the **Second Coming of Christ**—a time when Jesus would return in power and glory, the wicked would be judged, the righteous would be saved, and the **Kingdom of Heaven** would be established on Earth. Many still wait for this as a literal, visible event—a moment when the sky will open and the divine will descend.

But for those walking the spiritual path through metaphysical eyes, there is a deeper, symbolic truth hidden in these verses. Through years of study and spiritual discernment, many now understand the Bible not as a historical or literal text, but as an **allegorical and metaphysical manual**—a sacred book encoded with spiritual principles, cosmic metaphors, and initiatory teachings that mirror the **human soul's journey through density, division, and ultimately, divine remembrance.**

> The Second Coming is not one man descending from the clouds—
> it is **Christ Consciousness** rising from within.

The Kingdom of Heaven as the New Earth Frequency

In this expanded interpretation, the return of Christ is the **rise of collective divine consciousness**—what mystics call the *Christ frequency*. This is the embodiment of:

- Unconditional love
- Radical forgiveness
- Divine truth
- Personal sovereignty
- Peace, abundance, and unity

When the Bible says the righteous will be resurrected and the wicked will perish, it can be seen as the **fall of the old paradigm**: the systems rooted in manipulation, greed, violence, and fear. These systems are collapsing. What is being resurrected is the **true divine self** within each of us.

The "Kingdom of God" is not coming **to** us—it is rising **through** us. A New Earth where everyone eats, everyone heals, everyone thrives—not just the few. It is the frequency of Eden, restored through spiritual awakening.

Jacob's Ladder and the Ascension of Consciousness

Biblical metaphors like **Jacob's Ladder** (Genesis 28:12) —a stairway between Earth and Heaven—can be interpreted as a symbol for the **kundalini energy**, or the **spinal path of consciousness** awakening from the root to the crown. It represents:

- The ascent through chakras (or levels of awareness)
- The embodiment of higher truths
- The union of human and divine within each of us

Climbing Jacob's Ladder is the same as **ascending through inner transformation**. Each rung is a shift in consciousness—a letting go of fear, ego, and division, and a return to love, light, and unity.

Jesus as the Archetype of Awakened Humanity

In this light, Jesus was never meant to be worshiped as an unreachable figure—He was the blueprint for **what we all can become**: the awakened soul, the healer, the rebel of unjust systems, the divine human. His return is not physical—it is **energetic**. And it is happening now.

As people all over the world awaken, ascend, and embody higher frequencies of love, service, and divine truth—they are becoming the **return of Christ in action**.

> "The kingdom of God is within you." – *Luke 17:21*

A Prophetic Convergence

Whether through Christian scripture, the Hopi tablets, the Dogon cosmology, the Eagle and Condor prophecy, or the ancestral whispers rising from every bloodline—the message is clear:

We were never waiting for someone else to save us.

We are the return. We are the prophecy fulfilled. And the sacred timeline is now.

CHAPTER 6: THE 'CLAIRS' AND ENERGETIC SENSORY PERCEPTION

"Your soul speaks in frequencies. Your body interprets them through sensation. Intuition is your native language—you're just remembering how to speak it again."

What Are the 'Clairs'?

In the spiritual and metaphysical world, "The Clairs" are the different ways we perceive subtle energy—**the language of spirit, soul, and Source.** Just like our physical senses help us interpret the material world, our Clair senses help us interpret the **energetic** and **spiritual** realms.

These gifts are not just for mystics or psychics. They are **innate soul senses**, available to all of us. Many of us have used them our entire lives but didn't have the language or validation to understand what we were experiencing.

These senses are amplified when we are:

- Grounded in the body and nervous system
- Tuning into emotional awareness
- Practicing spiritual alignment and self-trust
- Operating from intuition rather than logic alone

The Spiritual Sixth Sense

We often speak of five physical senses: sight, hearing, touch, taste, and smell. But beyond these is a more subtle, often overlooked system of **energetic perception**—a spiritual nervous system that allows us to **feel what cannot be seen, hear what is unspoken**, and **know what the mind cannot explain**.

These are the **clear channels of spiritual communication** that activate when we're connected to our higher self, Spirit, or the energetic field around us.

The Clairs: Your Divine Receivers

Clairvoyance (Clear Seeing)

This is the ability to see images, symbols, colors, visions, or even flashes of insight through the mind's eye. Often associated with the **third eye chakra**, clairvoyance may show up as:

- Seeing flashes of color, light, or shadow
- Receiving vivid imagery during meditation or prayer
- Visualizing a scene or person you've never met, but later confirming its truth
- "Seeing" energy fields or auras around people or places

Connected to the **pineal gland**, the seat of inner vision and higher consciousness. It's often experienced first in dreams and altered states—then begins to surface in waking life. Clairvoyants often "see" what's coming before it arrives—not just externally, but energetically.

Clairaudience (Clear Hearing)

This involves hearing messages, phrases, names, tones, or even music that isn't audible to others. These messages may come from Spirit, your guides, your ancestors, or your higher self.

Clairaudients often describe:

- Hearing your name called with no one around
- Receiving messages in your own voice but with deeper clarity or wisdom
- Hearing ringing, tones, or music with spiritual significance
- Being "spoken to" during dreams, meditation, or prayer

Clairaudience is subtle and often dismissed as imagination. But it is one of the most direct ways Spirit communicates. It is associated with the **throat chakra** and the expression of divine truth. This sense often activates in quiet or liminal states—during prayer, before sleep, or in meditation.

Clairsentience (Clear Feeling)

Clairsentience is the ability to feel the emotions, energy, or physical sensations of people, places, or spiritual beings.

Clairsentients are often empaths who can:

- Feel the emotions of others as if they are their own
- Pick up energetic "vibes" from rooms or people
- Sense when something "feels off" without evidence
- Experience emotional responses when Spirit is near
- Getting chills or goosebumps when hearing something true

- Feeling heavy or anxious around someone with dense energy
- Knowing when someone is lying, even without evidence
- Receiving a sudden "hit" of joy, grief, or pain that isn't yours

Clairsentience is closely connected to the **solar plexus** and **sacral chakras**—our centers for emotion, trust, and intuitive gut instincts. Clairsentience is deeply linked to **empaths**, **healers**, and those who carry heightened nervous system sensitivity. It requires emotional intelligence, boundaries, and somatic practices to stay balanced.

Claircognizance (Clear Knowing)

This is when you simply know something without any logical explanation. It's a sudden insight, an unshakable truth, or a powerful inner "yes" that comes from beyond rational thought.

This might look like:

- Knowing someone's intentions without evidence
- Suddenly having a solution or answer to something you've never studied
- Downloading insight while in the shower, meditating, or driving
- Speaking truthfully or prophetically without planning to

This gift is often misunderstood as overthinking, but it's not rooted in anxiety—it's **instantaneous, neutral, and precise.** Claircognizance often activates during **alignment and stillness**, and it requires deep self-trust. This sense often aligns with the **crown chakra**, and is common in those who carry roles as messengers, teachers, or spiritual leaders.

Clairalience & Clairgustance – (Clear Smelling & Clear Tasting)

The ability to smell scents that are not physically present, such as a deceased loved one's perfume, the scent of flowers, tobacco, smoke, or sacred oils. These spiritual scents are often confirmation of **presence or memory**.

The ability to spiritually taste substances without physical contact—often connected to ancestral memory or symbolic messages. This may arise during rituals or ancestral connection practices.

This might look like:

- Smelling your grandmother's perfume when no one is wearing it
- Tasting a certain food that reminds you of home or ritual
- Sensing an elemental or spiritual presence through scent or flavor

These senses connect us to the **memory centers** of the brain and are often used by ancestors and guides to validate their presence.

You've Probably Been Using Your Clairs All Along

Most people use at least one Clair sense naturally:

- The mother who feels her child is in danger before the phone rings
- The creative who sees an entire vision before they begin painting or writing
- The healer who senses where pain is stored in a client's body

- The intuitive who dreams of an event before it happens

These are not fantasies or coincidences—they are the **soul's native tools** for navigating life, healing others, and tuning into higher wisdom.

Energetic Sensitivity: You're Not "Too Much"—You're Tuned In

Many people dismiss or deny their spiritual gifts because they were never taught how to understand or honor them. Sensitives, empaths, and intuitives are often told they're "overreacting," "imagining things," or being "too sensitive."

But what if that "too much" is your **sacred antenna**?
What if your emotions, your sensory awareness, and your energetic perception are **divine tools of guidance**?

> *You don't need to become psychic—you already are.*
> You simply need to remember how to trust the signals.

Why These Gifts Are Awakening Now

As the frequency of the planet rises, more people are experiencing an activation of their clair senses. This is not a coincidence—it is part of the **collective awakening**.

These gifts awaken when:

- You go through spiritual initiation or awakening
- You experience trauma that splits the veil
- You begin meditating, fasting, or doing energy work
- You reconnect with your ancestral lineage or spiritual path

- You move through the Dark Night of the Soul and emerge with new clarity

These abilities are not given only to the gifted few—they are the **birthright of the awakened many**.

Honoring Your Sacred Senses

The clairs are not just "woo-woo" phenomena. They are:

- **Neurological pathways opening to higher intelligence**
- **Energetic attunements to the collective and divine field**
- **Sacred communications from the unseen realms**

To trust them is to trust yourself.
To honor them is to honor your soul's design.
To develop them is to walk in your spiritual authority.

Activation Reflection: Awaken the Sacred Senses

> **Close your eyes.**
> **Breathe into your third eye, your heart, and your gut.**
> **Visualize a sphere of light illuminating each energy center.**
> **Say aloud or within:**
> *I am open to divine communication.*
> *I receive with clarity, grace, and discernment.*
> *I trust my gifts.*
> *I awaken the sacred senses within me.*
> *I am guided, protected, and connected.*
> *And so it is.*

Extrasensory Perception (ESP): What They Didn't Tell Us

> *"What society labels as 'supernatural' is often just suppressed knowledge."*

Extrasensory Perception (ESP) refers to the ability to gather information beyond the known senses—perceiving data, emotion, or energetic fields **without physical contact**. It includes phenomena like:

- **Telepathy** (mind-to-mind communication)
- **Precognition** (knowing before it happens)
- **Psychometry** (reading energy from objects)
- **Remote viewing** (seeing distant or unseen targets)
- **Clair abilities** (perceiving through energetic senses)

For centuries, these gifts were dismissed by mainstream institutions as fantasy or delusion—while simultaneously being **closely monitored, studied, and weaponized** by the very systems that denied their existence.

The Government's Secret Interest in Psychic Abilities

While the public was taught to scoff at the idea of ESP, global intelligence agencies were quietly conducting extensive research into psychic phenomena. Most notably:

- **Project Stargate (U.S. Military/CIA)**: From the 1970s to the mid-1990s, the U.S. government invested millions in the training and development of **remote viewers**—psychics who could "see" across space and time using only the mind. Their work included:
    - Locating hostages

  - Uncovering enemy secrets
  - Tracking nuclear materials

- **Soviet Psychic Research**: During the Cold War, the USSR developed similar programs to cultivate psychics for military and espionage purposes, often using children with naturally heightened ESP.

- **MK-Ultra**: While more controversial, some elements of MK-Ultra involved testing how psychic abilities might be influenced by trauma, psychedelics, and altered states.

These programs were not based on belief—they were based on **results**. Documents declassified by the CIA confirm that **ESP is real enough to be studied, tested, and utilized**—just not openly taught to the masses.

The "Gifted Programs": Quiet Experiments in Public Schools

If you grew up in the 80s, 90s, or early 2000s, you might remember being pulled from your classroom for a "Gifted and Talented" program. These programs were framed as academic enrichment, but many who participated now recall experiences that were… different.

- Memory games with oddly specific patterns
- Symbol recognition, mirror reading, and nonverbal problem solving
- Frequency tone tests ("Can you hear this?")
- Logic puzzles that resembled remote viewing frameworks
- Questions about dreams, gut feelings, and instincts

And many of the children placed in these programs now identify as:

- **Mystics, empaths, and intuitives**
- **Neurodivergent** (ADHD, autism spectrum, synesthetic, dyslexic)
- Spiritually sensitive and energetically attuned adults

It is believed by some that these programs were early **screenings** for naturally occurring ESP and multidimensional perception. Whether formal or informal, **these children were being tracked** —not for academic reasons alone, but for how they interacted with information beyond logic.

> These weren't just gifted kids.
> They were **tuned-in children**. Old souls. Walk-ins. Starseeds. Souls carrying codes from other timelines.

Neurodivergence as Spiritual Sensitivity

What if **neurodivergence** is not a disorder—but a **different operating system**?

Many neurodivergent individuals experience heightened sensitivity to:

- Light, sound, vibration, and energy
- Truth and emotional authenticity
- Injustice, hypocrisy, and energetic imbalances

They often:

- Hear things others don't
- Process language or emotion in nonlinear ways
- Have intense focus in altered states or subjects of passion
- Remember dreams vividly or astral project naturally
- See connections and patterns others miss

Rather than pathologize this, many are beginning to recognize that **neurodivergence is a spiritual frequency**—a form of energetic perception that traditional systems weren't built to support, but mysticism and quantum spirituality can embrace.

> *"What you called 'defiance' was discernment.*
> *What you labeled 'distraction' was multi-dimensional awareness.*
> *What you medicated was my magic."*

The Truth is: Your Gifts Were Never the Problem

You were never too sensitive.
You were never "broken."
You were never imagining things.

You were **perceiving differently—sensing deeply**—and in a world designed to keep people in linear, obedient, distracted boxes, **that makes you a threat to the illusion**.

And now? The illusion is crumbling.

Activation Reflection: Remembering the Coded Child

> **Close your eyes. Visualize yourself as a child —sitting in class, staring out the window, daydreaming, sensing something more.**

Now walk toward that child. Kneel down and say aloud or within:
I see you.
You weren't strange—you were sacred.
You weren't wrong—you were remembering.
You weren't lost—you were encoded.
And now, I reclaim the power that was always mine.
And so it is.

Timeline of ESP Research and Military Involvement

Late 19th – Early 20th Century: Foundations of Parapsychology

- **1882**: Establishment of the Society for Psychical Research (SPR) in London, aiming to investigate paranormal phenomena.

- **1911**: Stanford University conducts one of the first academic studies on ESP under psychologist John Edgar Coover.

- **1930s**: J.B. Rhine and Louisa E. Rhine at Duke University pioneer systematic ESP research using Zener cards, introducing statistical methods to parapsychology.

1970s: Cold War and the Rise of Psychic Research

- **1972**: The CIA initiates **Project SCANATE** to explore remote viewing capabilities in response to reports of Soviet psychic research.

- **1973**: Stanford Research Institute (SRI) begins formal remote viewing experiments under physicists Russell Targ and Harold Puthoff.

- **1977**: The U.S. Army's Intelligence and Security Command (INSCOM) establishes **Project GRILL FLAME**, focusing on the potential military applications of remote viewing.

1980s: Expansion and Consolidation

- **1985**: The Defense Intelligence Agency (DIA) consolidates various psychic research programs under **Project SUN STREAK**, continuing the exploration of remote viewing.

- **1986**: The program is renamed **Project CENTER LANE**, reflecting ongoing efforts to integrate psychic phenomena into intelligence operations.

- **1987**: The project undergoes another rebranding as **Project STARGATE**, aiming to centralize and streamline research activities.

1990s: Evaluation and Termination

- **1991**: Responsibility for the STARGATE program shifts to the Science Applications International Corporation (SAIC) for further research and evaluation.

- **1995**: The CIA declassifies and terminates the STARGATE program after a review concludes that remote viewing has not been proven effective for intelligence purposes.

2000s – Present: Renewed Interest and Modern Research

- **2014**: The Office of Naval Research launches a $3.85 million program to study "sensemaking," exploring intuitive decision-making and its potential applications in military contexts.

- **2025**: Declassified documents and renewed

public interest bring attention to historical ESP research, prompting discussions on the potential of human consciousness and perception

From Classified to Conscious: Bridging the Past and Present

"The gifts they studied in secrecy, we now reclaim in sovereignty."

The military and intelligence communities spent decades investigating extrasensory perception (ESP) under projects like **STARGATE**, not because they believed it was fantasy—but because **it worked**. Remote viewers accurately described secret facilities, missing persons, and enemy movements with nothing but the power of focused consciousness.

Now, as those projects have been declassified, their findings are re-emerging in the public realm—not as "espionage tools," but as **spiritual technologies**. The same methods used in government programs are now seen in modern practices like:

Remote Viewing → Intuitive Visualization & Akashic Reading

The controlled "mind journeys" used in the STARGATE program are now common in:

- **Guided Akashic Records readings**
- **Intuitive visualizations**
- **Quantum healing hypnosis (QHHT)**
- **Energy mapping and timeline jumping**

Remote viewing laid the groundwork for training the **clairvoyant muscle**—and today, it's taught in psychic development courses and used in quantum coaching containers to access future potentials or soul contracts.

Telepathy & Claircognizance → Quantum Communication

CIA reports explored **mental transmission of data** across space and time—what we now understand as **claircognizance** or **telepathic download**.

In today's consciousness practices:

- Lightworkers and channels receive messages from spirit guides, ancestors, and higher selves
- Empaths intuitively "know" truths about others without verbal exchange
- Healers tap into clients' emotional fields during sessions, often receiving symbolic or ancestral information without prior knowledge

These mirror exactly what intelligence officers were taught—**except now it's used for healing, not war.**

Altered States of Consciousness → Meditation, Lucid Dreaming & Psychedelic Integration

The government's interest in **altered states** through hypnosis, meditation, and psychedelics has found modern expression in:

- **Brainwave entrainment** (Theta, Delta for intuitive access)
- **Lucid dreaming techniques** used to access higher guidance
- **Somatic meditation and breathwork** to activate dormant "clairs"

- **Psychedelic-assisted therapy** (psilocybin, MDMA, ayahuasca) now being explored for spiritual breakthroughs and trauma healing

What they once saw as a tactical advantage is now seen as a **soul technology**—a way to return to our higher self.

ESP Training → Intuition Development & Energy Work

Programs that once trained psychics in secret are now mirrored in the rise of:

- Intuitive development schools
- Reiki and energy medicine certifications
- Quantum healing programs
- Multidimensional coaching and mentorship containers
- Somatic experiencing that activates the "body as oracle"

The exact skills they tried to isolate in labs are now being **reintegrated into sacred and ancestral practices**—just as the ancestors prophesied.

Why It Matters Now

We are at a point in collective consciousness where **everything once hidden is returning to the surface**. These intuitive abilities were never rare—they were suppressed. Controlled. Weaponized.

And now?
We're reclaiming them for **truth, healing, and collective liberation**.

The fact that you can now:

- Meditate and access visions
- Sense a person's energy from across the world
- Know when something is about to happen
- Receive guidance in dreams or downloads

...is not coincidence. It's **consciousness remembering itself.**

You Are the New Researcher

Where the government once experimented behind closed doors, you are now the **sovereign scientist** of your own mind, energy, and spiritual gifts. You don't need a lab coat or a clearance badge —you need only presence, practice, and permission to trust what you already know.

Activation Reflection: From Secret to Sacred

Place your hands on your third eye and your heart.
Breathe deeply. Remember.

Now speak aloud or within:
I reclaim the knowledge that was hidden.
I remember what they tried to suppress.
I receive my intuition as divine intelligence.
My consciousness is my power.
My perception is my liberation.
And so it is.

CHAPTER 7: DREAMS, VISIONS, AND DIVINE DOWNLOADS

"The dream world is not an escape—it is a return. A place where the soul remembers what the mind forgets."

Your Dreams Are Not Random—They're Revelations

In every ancient culture, dreams were revered as messages from the spirit realm. Pharaohs consulted dream interpreters before making decisions. Indigenous medicine people received healing songs in visions. Prophets across traditions encountered divine wisdom while asleep.

Yet in modern times, we've been taught to dismiss our dreams as meaningless—just neural activity, stress processing, or scattered thoughts.

But the spiritual truth is this: **Your dreams are sacred communications** from your higher self, your ancestors, your guides, and the universal field of divine intelligence.

Dreams are where the **veil is thinnest**—where time collapses, identities dissolve, and truth rises in metaphor, symbol, and sensation.

The Three Realms of Soul Communication

There are three primary ways divine information flows through us beyond waking logic:

Dreams

Accessed in altered states (sleep, deep rest, REM), dreams allow you to:

- Rehearse future timelines

- Release stuck energy or trauma
- Visit spirit guides, ancestors, or past lives
- Receive symbols, metaphors, and messages to decode

Some dreams are nonsensical, yes. But many are deeply encoded blueprints waiting to be remembered.

> *If you've ever dreamed of flying, falling, running through water, losing teeth, or being chased—your soul was speaking to you.*

Visions

Unlike dreams, visions occur in waking states. These include:

- Sudden inner imagery during meditation or prayer
- Flash-forwards (prophetic glimpses of what's coming)
- Symbolic downloads that appear behind closed eyes
- Momentary shifts where time slows or clarity sharpens

Visions may come in silence, stillness, or even chaos. They're not "seen" with the physical eye, but with the **inner eye**—your **third eye**, the seat of psychic and divine vision.

Divine Downloads

These are instant, often undeniable floods of **knowing**, **feeling**, or **direction** that arrive out of nowhere—beyond logic or study. Downloads may come:

- In the shower
- Just as you're waking

- During meditation
- While writing, creating, or speaking

They feel like lightning strikes of **truth**. You didn't "figure it out"—you *received* it.

> *A download is the soul delivering wisdom the ego was not ready to seek, but the spirit was ready to speak.*

The Symbolic Language of Spirit

Dreams, visions, and downloads don't always come with clear instructions. They speak in **symbol, story, and feeling**. Learning to **interpret** them is an art of returning to your intuitive language.

Common Symbols & Their Meanings:

- **Water** – emotions, healing, cleansing, spiritual rebirth
- **Teeth** – power, communication, growth, anxiety
- **Flying** – freedom, transcendence, spiritual elevation
- **Chase** – avoidance, fear, confrontation of truth
- **Snakes** – transformation, kundalini energy, rebirth
- **Death** – endings, ego death, transformation (rarely literal)

Your symbols may vary by cultural background, spiritual lineage, or personal experience—**and that's the point**. Your dream language is unique to *you*. You don't need a textbook—you need to trust what your body and spirit feel when the dream returns.

Downloads and Prophecy: Not Just for "Special People"

We've been taught that visions and prophecy are only for saints, priests, or ancient prophets. But in truth, **everyone receives divine insight.** The more you open, the more clearly it flows.

- That gut feeling? A download.
- That dream that felt *too real*? A visitation.
- That idea that came through you like lightning? Divine intelligence moving through your system.

You're not imagining things. You're remembering how to listen.

Why You Might Not Remember Your Dreams

If your dream recall feels blocked, it's not that you don't dream—it's that your spirit has been **trained not to remember**. This can happen from:

- Emotional suppression
- Fear of what might be revealed
- Overactive logical thinking
- Nervous system dysregulation (hypervigilance blocks reception)
- Lack of spiritual safety or grounding

To reconnect:

- Create a dream altar or keep a notebook beside your bed

- Set an intention before sleep: "I am ready to receive and remember"
- Practice morning stillness—don't grab your phone right away
- Drink water and move slowly upon waking
- Reflect, even if all you remember is a color, feeling, or word

Activation Reflection: Becoming a Dream Receiver

Lay or sit with your eyes closed.
Visualize a golden thread connecting your third eye to the sky.
Feel your body resting in a cocoon of soft light.

Whisper aloud or within:
I am open to receive divine dreams.
I remember what I am meant to remember.
I honor the symbols, feelings, and messages of the night.
My dreams are sacred conversations with my soul.
And so it is.

Lucid Dreaming & Astral Travel: The Conscious Soul Explorer

"Dreams are where we remember who we are.
Lucid dreams are where we choose who we become."

Not all dreams are passive experiences. Some are conscious, intentional, and utterly transformational.

Lucid dreaming and **astral travel** represent the soul's ability to navigate the quantum and spiritual realms **with awareness**—to participate, observe, and even interact with the non-physical dimensions of existence.

They are not fantasy. They are not hallucination.
They are **advanced states of consciousness** that allow you to:

- Visit higher dimensions
- Interact with spirit guides, ancestors, and divine beings
- Heal trauma and rewrite subconscious programming
- Receive profound insight and cosmic guidance
- Explore your soul's origin and purpose

Lucid Dreaming: Awakening Within the Dream

Lucid dreaming occurs when you **become aware** that you are dreaming while still inside the dream. This awareness allows you to:

- Consciously direct your dream experience
- Ask questions of the subconscious or higher self
- Fly, shape-shift, or travel to desired places or time periods
- Practice spiritual rituals or energy work
- Rehearse future events and explore alternate timelines

Lucid dreaming is a doorway to the Akashic Field—the universal record of all existence—and it's one of the most accessible forms of conscious soul travel.

How to Cultivate Lucid Dreams:

- **Reality Checks**: Ask yourself throughout the day, "Am I dreaming?"—this habit bleeds into dream space.

- **Dream Journaling**: Writing dreams consistently trains the mind to remember and stabilize lucidity.
- **Set Intentions**: Before sleep, say: "Tonight, I will realize I am dreaming."
- **Wake-Back-to-Bed Method**: Wake up after 4–6 hours of sleep, stay awake briefly, then return to bed focused on re-entering a lucid state.

Lucid dreaming is the art of reclaiming awareness in the most intimate part of your unconscious world.

Astral Travel: The Spirit's Journey Beyond the Body

Astral travel (also known as astral projection or soul travel) occurs when your **consciousness separates from the physical body** and explores the astral plane—a multidimensional space that exists between the Earth realm and higher spiritual realms.

Unlike lucid dreams, which take place within the mind's dreamscape, astral travel takes you into **real energetic dimensions** where you can:

- Visit people or places across time and space
- Meet ancestors or spirit guides
- Observe past lives
- Explore other planets, soul realms, or timelines
- Access cosmic libraries of divine information

Astral projection is natural—it often happens during near-death experiences, deep meditation, and spontaneous dream journeys. You don't force it—you allow it.

Signs You're Astral Traveling:

- Feeling a vibrational "buzz" or intense frequency before sleep
- Hearing ringing, rushing, or popping sounds as your spirit loosens
- Seeing your body from above (out-of-body experience)
- Traveling through tunnels or light portals
- Waking up with knowledge or imagery you didn't previously have

How to Prepare for Astral Travel

- **Ground first**: Begin with breathwork and intention to remain energetically safe and balanced.
- **Create a ritual**: Light a candle, play 432 Hz music, or speak a protection prayer.
- **Call in your guides**: Ask your spirit team to accompany you for guidance and protection.
- **Use visualization**: Imagine your spirit body rising out of your physical form gently and effortlessly.
- **Stay in trust**: Let go of fear—your body will always call you back when it's time.

The Purpose of Conscious Dream Travel

Lucid dreaming and astral projection are more than curiosities—they are **soul technologies** that allow you to:

- Retrieve pieces of your divine identity
- Heal fragmented parts of your spirit
- Practice spiritual gifts before anchoring them in waking life
- Understand multidimensionality from *experience*, not theory
- Deepen your relationship with Source

These practices are invitations to walk between worlds, to become fluent in the symbolic language of your higher self, and to activate your **cosmic remembrance**.

Activation Reflection: Conscious Journeying

> **Before sleep, sit with intention.**
> **Say aloud or within:**
> *I am safe to travel the sacred realms.*
> *I remember my journey through dreams and spirit.*
> *I awaken within the dream.*
> *I trust where my soul is guided.*
> *I walk the invisible path with divine protection.*
> And so it is.

CHAPTER 8: MULTIDIMENSIONALITY, TIMELINES, AND QUANTUM IDENTITY

"You are not a single being with a linear life.
You are a cosmic soul living multiple lives in one breath."

You Are a Multidimensional Being

"You are not limited by the skin you're in or the story you were born into.
You are the cosmos in human form—remembering itself one layer at a time."

From the time we're born, we're conditioned to see ourselves as singular—one body, one name, one lifetime, one role. But your soul knows better.

In truth, you are:

- Existing in multiple dimensions simultaneously
- Expressing infinite versions of yourself across parallel timelines
- Living both within this moment and beyond it
- Shifting and shaping reality with every decision, thought, and vibration

You are **quantum**. That means you are not fixed—you are fluid. You are not static—you are infinite potential made manifest.

> This isn't philosophy. This is physics.
> And your soul already knows how to move between the layers.

Scientific Perspective: Dimensions Beyond the Visible

In physics and cosmology, the idea that we live in a **multidimensional universe** is well established—especially through **String Theory** and **M-Theory**, which propose that beyond the three physical dimensions and time (the 4th dimension), there are **10, 11, or even more dimensions** woven into the fabric of reality.

- According to String Theory, **subatomic particles** are not point-like, but are vibrating strings that operate across **multiple dimensions**.

- These dimensions are not always "out there"—they are **folded into our reality**, like frequencies we don't see but are still surrounded by, much like **radio waves** or **infrared light**.

- Quantum physicist **David Bohm** spoke of the *Implicate Order*—a hidden dimension where all things are connected, and reality is **holographic** and interconnected beyond time and space.

What does this mean for us as humans?

We're not just interacting with a linear, cause-and-effect world. We're living in a **field of infinite potential**, where consciousness affects reality and identity is not limited to one place in time.

Neuroscience and Consciousness Studies

While mainstream neuroscience often restricts consciousness to brain function, pioneers in the field are exploring new paradigms:

- **Dr. Stuart Hameroff** and **Sir Roger Penrose's Orch-OR**

Theory proposes that **quantum processes** in the brain's microtubules may allow for **non-local consciousness**, connecting us beyond physical limitations.

- **Dr. Eben Alexander**, a neurosurgeon who had a near-death experience (NDE), documented consciousness beyond the brain in his book *Proof of Heaven*. He describes existing in a multidimensional realm of pure awareness and light—without any neurological activity in his brain.

These insights suggest that **consciousness is not local to the body**—and if that's true, then your consciousness can exist in multiple realities at once. You are not bound to this single form—you are accessing data, emotions, and experiences **from beyond it.**

Spiritual and Mystical Traditions: The Soul Across Realms

In the mystical and esoteric traditions, multidimensionality is a well-known truth:

Hinduism

The soul (*Atman*) is said to exist beyond time and reincarnates through many lives. The concept of the **Koshas** speaks to layers of being:

- Physical (Annamaya)
- Energy (Pranamaya)
- Mind (Manomaya)
- Intuition (Vijnanamaya)
- Bliss (Anandamaya)

These are **dimensional aspects of Self**, each one more subtle and connected to Source.

Mystical Christianity

Mystics such as **Teresa of Ávila** and **Meister Eckhart** described the soul as a **multi-chambered castle**, where divine union occurs in stages of awakening. This reflects **levels of spiritual dimension** or consciousness.

Indigenous & Shamanic Cosmology

Many Indigenous and African spiritual traditions view the human being as connected to:

- **Earth realm** (physical life)
- **Spirit realm** (ancestors, guides, nature spirits)
- **Sky realm** (Source, future ancestors, star beings)

The soul moves through these planes, not just in ceremony—but in sleep, trance, prayer, and death. The soul is never "just here"—it is **constantly moving**, **learning**, and **remembering** across time and space.

Energetic Evidence: The Aura and Light Body

Modern devices like **Kirlian photography** and **biofield imaging** have visually confirmed what mystics and healers have long known:

- The body emits an **energy field** (also called the **aura**)
- This field extends beyond the skin and fluctuates with **thoughts, emotions, and spiritual practices**
- The **chakras**, or energy centers, serve as multidimensional portals within the body

When these energy fields shift, so too does your **vibration**, your **perception**, and your **alignment with different dimensions** of reality.

This is how you can:

- Wake up from a dream with wisdom you didn't have before
- Meet someone and "know" them instantly
- Change your thoughts and suddenly shift your entire life's path
- Feel a deep grief that seems older than this lifetime

These are not random experiences. They are signs of your **multidimensional self waking up**.

You Are the Bridge Between Worlds

You are not just a person living a linear life.
You are a **vibrational being navigating layers of reality**:

- Your body exists in 3D.
- Your thoughts operate in 4D.
- Your soul dances in 5D and beyond.
- You receive dreams from one timeline and walk through another.
- Your emotions carry the echoes of lives you may not remember but deeply feel.

When you recognize your multidimensionality, you reclaim your **spiritual sovereignty**. You no longer identify with only your traumas, your job title, your age, or your ancestry—you remember that you are:

A time traveler
A dream walker
A consciousness explorer
A divine spark woven through infinite dimensions

Activation Reflection: Remembering the Multidimensional Self

> **Place both hands on your heart and close your eyes. Inhale deeply and visualize yourself expanding beyond your body—feel your energy ripple through layers of light.**
> **Say aloud or within:**
> *I am more than one lifetime.*
> *I am more than one identity.*
> *I am infinite in expression.*
> *I walk between worlds.*
> *I trust the wisdom of my multidimensional soul.*
> And so it is.

What Is Multidimensionality?

Multidimensionality is the ability to perceive and operate across multiple layers of reality—not just the material (3D), but also the:

- **4D** (thoughts, time, emotion, dreams)
- **5D** (unconditional love, unity consciousness, soul truth)
- **6D+** (akashic awareness, archetypal energy, divine blueprinting)

As you raise your frequency, you begin to:

- Remember your soul gifts and missions
- Access divine guidance without external validation
- Feel connected to past lives, future selves, and parallel versions of "you"
- Shift timelines, identities, and realities in real-time

What Are Timelines and How Do They Shift?

> *"Every version of you already exists.*
> *The you who healed. The you who soared. The you who stayed. The you who said no.*
> *Shifting timelines is about choosing which version you're ready to meet."*

In quantum science, every possibility you could ever live **already exists**. Every choice you could make creates a new timeline—an alternate version of reality branching from the now.

Timelines are not linear—they are layered. You move between them based on:

- Your frequency (emotions, beliefs, energy)
- Your choices and aligned actions
- Your ability to release karmic loops and outdated identities

You can shift timelines by:

- Making a different decision
- Changing your internal belief system

- Releasing trauma that kept you locked in a past pattern
- Aligning with your soul's highest truth

This means the "future" isn't waiting to happen—it's responding to who you become *now*.

> *Timelines don't just unfold—you activate them.*
> *You are not on the ride—you are the driver.*

Quantum Physics and Parallel Possibilities

In **quantum physics** time is **nonlinear**, it is flexible, and multidimensional.

- The **Many-Worlds Interpretation** of quantum mechanics (Hugh Everett, 1957) suggests that every time you make a decision, the universe **splits**, and an alternate timeline is created where you made the *other* choice.
- The **Double Slit Experiment** showed that particles behave differently when observed, meaning **consciousness affects matter**—what you expect to happen actually **shapes the outcome**.
- In essence, reality is not fixed—it's a fluid web of **possibilities that respond to your vibration, your choices, and your level of awareness**.

So what are timelines?
They are **energetic storylines** that play out based on who you're being and what you're believing at any given moment.

Timelines as Frequencies, Not Destinations

Think of timelines like **radio stations**. Every station is already broadcasting—each one with a different song, a different vibe, a different story.

- If you're tuned into 88.3 FM, you're not hearing what's on 101.9 FM—even though it's *still playing.*

- In the same way, if you're living in fear, doubt, or scarcity, you're "tuned into" a timeline that reflects that frequency: delays, obstacles, disconnection.

- But when you shift your energy—your thoughts, beliefs, emotions—you shift frequencies. And therefore, you shift the **timeline** you're aligned with.

You didn't "force" a new life.
You simply **tuned into the version of you that already exists in a better one.**

How Do Timelines Shift?

Timeline shifting happens both **subtly** and **suddenly**. Sometimes it's a gradual unfolding, other times it feels like the world rearranged overnight.

You shift timelines when you:

- **Make a new choice** from a healed perspective

- **Release old patterns** that were keeping you anchored in a loop

- **Let go of people or environments** that no longer match your vibration

- **Receive an insight or download** that rewires your understanding of life

- **Choose courage over comfort**, alignment over fear

- **Activate forgiveness**, gratitude, or divine trust—

instantly recalibrating your field

In spiritual terms, this is **quantum leaping**—a rapid shift from one state of reality to another, often without linear explanation.

Signs You've Shifted Timelines:

- You feel "lighter," like a weight was lifted without external change
- You wake up one day and something just *clicks*
- Opportunities begin appearing that felt previously blocked
- People fall away or new aligned ones enter your life
- You feel like you're in a new version of your old life—same place, but different frequency
- You *barely recognize* who you were even a month ago

Sometimes, these shifts are subtle—other times they are cosmic events: **a breakup, a breakdown, a breakthrough, a revelation**. These are timeline crossroads where your soul says: *Choose again. Choose now. Choose higher.*

Higher Self Navigation & Timeline Consciousness

Your **Higher Self**—the eternal, ascended aspect of your soul—already exists in the highest possible timeline of your life. The more you align with this version of yourself, the more you collapse timelines and **merge with your divine potential**.

This is where intuition becomes your GPS:

- That "gut feeling" is your Higher Self nudging you toward alignment

- That inner nudge to change careers, move, or speak up is a timeline activation
- That moment you say *yes* to something unknown but deeply right? That's a leap

Timeline shifting doesn't require magic. It requires courage.
To choose the version of you who already remembers who they are.

Spiritual Integration: Working with Timeline Jumps

In mysticism, timeline shifts often show up as:

- **Déjà vu** or feeling like you've "lived this before"
- **Repeating dreams** or messages from the future
- **Energetic whiplash** (feeling disoriented, emotional purging, rapid realizations)
- **Synchronicity storms** (1111s, unexpected meetings, divine detours)
- Feeling like **you're being "tested" or initiated** into a new level of self

These are all signs that your reality is **rearranging around your evolution.**

Timeline Sovereignty: You're Not Stuck, You're Choosing

You're not trapped in your past. You're not doomed by your mistakes.

You are always one decision, one breath, one alignment away from a new reality.

There is a version of you who already:

- Wrote the book
- Healed the wound
- Left the job
- Spoke their truth
- Forgave the unforgivable
- Became who they were born to be

You don't need to become her—you need to remember her.
Step into her. And walk as her. Now.

Activation Reflection: Shift the Timeline

Sit in stillness.
Breathe into your belly.
Visualize a path branching into two or more directions in front of you.
See your future self—the one aligned with your highest timeline—standing at the end of one path, glowing with clarity and purpose.

Ask:
"What decision must I make today to meet you there?"

Affirm:
I am not bound to the past.

I shift timelines with intention.
I collapse fear and expand into truth.
I choose the highest path for my soul.
I walk into alignment with who I came here to be.
And so it is.

You Exist in Many Versions, Simultaneously

In the quantum field, everything that **can happen, does happen** —across infinite realities, timelines, and dimensions. Every time you make a choice, a new version of you is created that made the opposite one. These are called **parallel selves** or **alternate timelines**.

Simultaneously, your soul is **eternal**—reincarnating through different bodies, eras, identities, and experiences. These are known as your **past lives**, although in quantum time, they're more accurately described as **simultaneous soul expressions**.

What this means is that:

- There are versions of *you* who exist in **this lifetime** that made different choices.
- There are versions of *you* from **other lifetimes** who lived entirely different stories—yet their soul essence is still **part of your energetic DNA**.
- These selves are not separate from you—they are **YOU**, just across the veil.

And as you raise your frequency, awaken your gifts, or walk into new versions of yourself, you begin to **merge with these parallel and past aspects**, collapsing timelines and integrating their wisdom, power, and memory.

Signs You're Merging with a Parallel or Past Self

You may feel:

- A sudden personality shift or new desires that feel unfamiliar
- Intense emotional waves that don't match your current circumstances
- Vivid dreams or flashbacks that feel like memories (but aren't from this life)
- Recurring fascinations with certain eras, cultures, or names
- A deep fear, wound, or gift you've had "your whole life" with no known cause
- A voice in your head or energy in your body that feels like a wise version of *you* guiding you

These are not hallucinations. They're integrations.

Your soul is **collapsing karmic fragmentation**—bringing you back into wholeness by reuniting with the versions of you who have what you now need:

- Courage from your warrior lifetimes
- Intuition from your priestess selves
- Discernment from your past mistakes
- Sacred rage from lives where you were silenced
- Forgiveness from those who once caused harm

This is soul alchemy: reclaiming the fragmented pieces of self and becoming whole in the now.

How and Why These Mergings Happen

Parallel and past self-merging usually occurs during:

- **Spiritual awakenings** or "ego deaths"
- Major life transitions (divorce, loss, move, career shift)
- Deep healing work (therapy, plant medicine, shadow integration)
- Quantum leaps or frequency upgrades
- Interactions with soulmates or karmic mirrors who activate ancient memories
- Visiting sacred lands that stir a remembrance

These events act like **activators**, calling forth the energy of a specific aspect of your soul that holds the *codes* for your next evolution.

Merging is not always gentle—it can feel like an identity crisis, a deep grief, or an emotional storm. But on the other side is **embodiment**—you begin to feel **more you than you've ever felt**.

The Spiritual Purpose: Soul Remembrance and Timeline Collapse

As you merge with parallel and past versions of yourself, you:

- Reclaim lost power and integrate soul gifts
- Resolve karmic loops by offering healing, compassion, and closure

- Collapse timelines that are no longer relevant or aligned
- Accelerate your spiritual growth by merging lessons instead of repeating them
- Increase your intuitive and energetic bandwidth

This is how you walk as your **Higher Self** in the now.
You are no longer just "you"—you are a **collective of soul wisdom**, braided into one timeline.

Practices to Support Integration

1. **Mirror Gazing** – Look into your own eyes and ask: "Who else am I?" Let the image shift. Speak to the versions of you who come forward.
2. **Past Life Regression** – Through guided meditation, hypnotherapy, or intuitive journaling, access previous lifetimes for healing and insight.
3. **Timeline Journaling** – Write letters from your future or past self. Ask: *What do you know that I need right now?*
4. **Somatic Tracking** – Notice where in your body you feel shifts or sensations. Past life trauma often lives in the tissues.
5. **Soul Retrieval Ceremonies** – Work with a trusted healer or guide to call back fragmented pieces of your soul from past or parallel lives.

You Are the Living Portal

Every time you heal, forgive, speak truth, or rise—you are not just doing it for yourself.

You are liberating:

- The version of you who couldn't
- The lifetimes where you were silenced
- The timelines that ended in tragedy
- The selves who prayed for someone like *you* to remember

> *You are the reincarnation of every version who ever dreamed of becoming free.*
> *And now... you are the one who sets them all free.*

Activation Reflection: Integration of the Many Selves

> **Close your eyes and place one hand on your heart, one on your lower belly.**
> **Visualize thousands of stars—each one a different version of you—gently moving toward your body like light codes.**
>
> **Whisper aloud or within:**
> *I call back all of me.*
> *The me who loved.*
> *The me who lost.*
> *The me who led.*
> *The me who fell.*
> *The me who remembered.*
> *I merge in love. I merge in power. I merge in wholeness.*
> *I am One. I am infinite. I am now.*
> And so it is.

Quantum Identity: The Soul Beyond the Self

Your "identity" is not your name, your job, your trauma, or your past.

Your **true identity** is your **soul frequency**—the energetic signature that transcends every role, label, or lifetime.

When you release attachment to the old self, you open to:

- Rapid transformation
- Timeline acceleration
- Soul-aligned relationships and mission
- Divine downloads and multidimensional perception

This is what's meant by "quantum leaping"—you shift into a **higher expression of your being**, where what once took years now manifests in weeks or days.

> *You are not reinventing yourself.*
> *You are remembering who you were before the world told you who to be.*

Practical Ways to Embody Your Multidimensional Self

1. **Energetic Attunement** – Use breath, sound, or meditation to raise your vibration and tune into higher timelines.
2. **Identity Release** – Let go of old labels and roles that no longer reflect who you are becoming.
3. **Mirror Work** – Speak directly to your future or parallel self in the mirror, asking what they know that you don't yet.
4. **Timeline Tracking** – Journal moments where your reality shifted—new opportunities, people, or

insights that appeared "out of nowhere."

5. **Quantum Decision-Making** – Ask: *Does this align with the highest version of me?* before acting.

Activation Reflection: I Am Multidimensional

Close your eyes.
Visualize every version of you—past, future, parallel—gathering in a circle of light.
Breathe deeply and allow them to merge into your body, your being, your now.

Say aloud or within:
I am a multidimensional being.
I exist across timelines and dimensions.
I choose to align with my highest self.
I embody all that I've ever been and all I'm becoming.
I am sovereign. I am infinite. I am divine.
And so it is.

CHAPTER 9: EMBODIMENT OF THE DIVINE HUMAN

*"You are not here to escape your humanity.
You are here to make it divine."*

What Is Embodiment?

Embodiment is the **full integration** of spiritual awareness into your lived experience. It's the process of **bringing your soul into your body**, not just conceptually—but physically, emotionally, energetically, and behaviorally.

To embody the Divine Human means:

- Living in alignment with your soul, moment by moment
- Expressing love, truth, and compassion *as a default, not an exception*
- Operating with clarity, purpose, and discernment—without ego
- Letting your nervous system, body, and words become **vehicles of spirit**
- Merging your mystical knowing with grounded presence

This is not about becoming perfect.
This is about becoming *fully integrated.*

> *It's not how much light you can talk about.*
> *It's how much light you can* ***hold*** *in your everyday life.*

The Science of Embodiment: Neurobiology and Integration

Embodiment is supported by **applied neuroscience**, trauma studies, and somatic psychology:

- **The vagus nerve**, which connects the brain to the body, regulates **nervous system safety**—a key aspect of embodiment. The more regulated your nervous system, the more divine frequencies you can hold.

- **Neuroplasticity** allows you to rewire your reactions, emotions, and behaviors through repetition, awareness, and alignment.

- Trauma healing (via modalities like Somatic Experiencing, EMDR, or EFT) creates space for the body to release old imprints and **invite in new soul-aligned behaviors**.

Your **biology is not separate from your spirituality**—it's the channel through which your divine consciousness is expressed. Your body is the **altar**. Your nervous system is the **temple**.

Spiritual Truth: You Are the Living Bridge

All the ancient teachings—Kundalini in yoga, Christ Consciousness in Christianity, Ka and Ba in Kemetic philosophy —speak to this one truth:

> You are the **meeting point** between heaven and earth.
> You are the **incarnated divine**.

This means:

- Your emotions are sacred messengers

- Your body is an oracle
- Your pleasure is power
- Your presence is medicine
- Your voice is activation
- Your alignment is service

When you embody your divine self, you no longer chase miracles.
You **become the miracle.**

From Awakening to Embodiment

Many people awaken to their divine nature but struggle to embody it. That's because awakening is *expansion*, but embodiment is **integration**.

Awakening	**Embodiment**
Receiving visions, downloads	Taking inspired, aligned action
Feeling bliss in meditation	Holding that peace in real conflict
Realizing you're a healer	Living as a healer, daily
Knowing your soul path	Walking it, even when it's hard
Channeling truth	**Becoming** that truth

Awakening opens the door.
Embodiment walks through it.

Practices to Support Divine Embodiment

1. **Somatic Presence:** Spend time every day *in your body*. Breathe, stretch, feel, dance, touch the earth.
2. **Soulful Boundaries:** Say "no" to what contracts you and "yes" to what expands you. Integrity is sacred.
3. **Heart Coherence:** Practice aligning your breath with your heart rate. This anchors you in love.
4. **Sacred Speech:** Use your words as spells—intentional, clear, and kind. Your voice carries divine vibration.
5. **Aligned Action:** Follow through on soul nudges, no matter how small. Every act is an activation.

The Result: Sovereign, Rooted, Radiant Humanity

When you embody the Divine Human, you:

- Walk in rooms and change the frequency
- Speak, and people feel the transmission behind the words
- Create with ease because you're in alignment, not struggle
- Love without losing yourself
- Lead without needing control
- Heal others simply by *being* who you are

This is the new earth template:
Healed, embodied, fully expressed humans who are both divine messengers and divine messes—unafraid to live, love, cry, roar, build, rest, and rise... all as sacred.

> *You are not here to escape the world.*
> *You are here to remake it—through the vibration of your embodied truth.*

Activation Reflection: I Am the Divine in Form

> **Sit or stand tall. Place your feet flat on the earth.**
> **Roll your shoulders back, lift your head.**
> **Place one hand on your heart, one on your lower belly.**
> **Breathe into your center.**
>
> **And speak aloud or within:**
> *I am the soul made flesh.*
> *I am the divine expressing through human form.*
> *I release the need to be perfect—I choose to be whole.*
> *My body is sacred. My breath is prayer.*
> *I walk in truth, love, power, and peace.*
> *I am the bridge between worlds.*
> *I am the embodiment of the divine.*
> And so it is.

CHAPTER 10: THE HEALING POWER OF INTEGRATION

"Healing is not about fixing what is broken—it's about remembering you were whole all along."

Integration Is the Bridge Between Awareness and Wholeness

Integration is the process of taking the **awakened insights**, the **soul-level revelations**, and the **embodied experiences**—and weaving them into your everyday life, your identity, and your nervous system.

Where awakening is the flash of insight, **integration is the grounding**.
Where embodiment is the expression, **integration is the recalibration**.
It's the sacred stitching together of your spiritual, emotional, physical, and mental self into **one unified field of being**.

Why Integration Is Essential for Healing

Many people seek healing through:

- Accumulating knowledge
- Repeating rituals
- Escaping pain through spiritual bypassing
- Endlessly chasing the "next level"

But healing doesn't happen through *more*.
Healing happens when we **pause**, feel, and allow all parts of

ourselves to come home.

Integration is the part of the journey where we:

- Stop running from the shadows and **invite them to speak**
- Bring compassion to our past, **not shame**
- Acknowledge where the body still holds pain, **and gently release it**
- Let the nervous system **catch up to the spirit's expansion**

You don't need more magic.
You need time to absorb the magic you've already called in.

The Science of Integration: Neurobiology and Regulation

The process of integration is deeply supported by the **brain's natural wiring for coherence and homeostasis**:

- **Neuroplasticity** allows the brain to create new neural pathways when we choose a new thought, response, or behavior—**but only through repetition, rest, and regulation**.
- The **limbic system**, which stores trauma and emotional memory, responds to **safety, stillness, and self-compassion**, not logic or force.
- **Somatic integration**—through movement, breath, sound, and sensory awareness—allows the body to **release trapped emotion** and **reprogram the stress response**.

This is why spiritual growth must include nervous system work.

Without integration, awakening can overwhelm the system. With it, we experience **sustainable transformation**.

Combining Spirituality and Science in Holistic Practices

The healing modalities of today are no longer divided into "Western" vs. "Eastern," "clinical" vs. "intuitive," or "science" vs. "spirit." More and more practitioners are realizing that the most powerful healing happens when we **blend the wisdom of the unseen with the rigor of the observable**.

From Neuroscience to Inner Knowing

- **Neuroplasticity** proves that we can rewire trauma-based thoughts.
- **Epigenetics** shows that emotional environments influence our gene expression.
- **Somatic therapies** validate what Indigenous cultures have known for millennia: the body holds stories, and movement is medicine.

When we pair these findings with:

- Intuitive energy work
- Spiritual coaching
- Meditation, reiki, breathwork, and quantum healing

...we create a **full-spectrum healing experience** that treats the root, not just the symptom.

This is the **future of healthcare**—one that sees the human being as more than a machine to diagnose, but a soul to support, a story to honor, and an energy field to rebalance.

> *It's no longer science vs. spirit.*
> *It's science validating what spirit has always known.*

Client Stories: The Power of Integration in Action

> *(Names changed, HIPAA compliant, and shared with consent)*

Tasha's Story: From Chronic Pain to Emotional Release

Tasha had suffered from fibromyalgia for over 7 years. She had tried everything—medications, physical therapy, diet changes—but nothing lasted. During a holistic intake, we discovered that her pain flared most during emotional stress.

Through somatic therapy, breathwork, and visualization, we uncovered repressed trauma from her teenage years that she had never spoken about. Once she began gently integrating that part of her story—through movement, journaling, and ancestral healing—her physical pain began to subside. Within six months, she was medication-free and leading movement classes for women healing from trauma.

> "Once I stopped resisting my emotions and let them speak, my body finally stopped screaming."

Malik's Story: Awakening the Healer Within

Malik was a military veteran struggling with PTSD and anxiety. Traditional therapy helped, but something was missing. When he began energy healing and guided hypnotherapy alongside clinical support, he experienced profound dreams of ancestral warriors and visions that showed him a deeper calling.

With support, Malik integrated these messages, began training in herbalism, and started a nonprofit supporting other Black veterans with spiritual healing circles.

> "I didn't need to be cured—I needed to be connected.

My healing started when I realized I wasn't broken. I was disconnected from who I truly am."

The Future of Integrative, Quantum-Aligned Healthcare

We are entering a new era of healing—one that honors the **multidimensional human**.

The New Model of Care:

- Treats **trauma, not just symptoms**
- Combines **energy medicine with nervous system regulation**
- Uses **quantum diagnostics** to assess the biofield and emotional imprints
- Respects **spiritual awakening** as part of the healing journey—not a "mental illness"
- Incorporates **ancestral healing, cultural restoration, and epigenetic release**

This model understands that healing is not linear—it's **quantum**.

- It sees the client not just as a patient, but as a **sovereign soul**.
- It acknowledges that we are all **storytellers**, **energy fields**, and **divine blueprints** in motion.
- It replaces the old paradigm of "fixing the broken" with a new one: **awakening the whole**.

The medicine of the future is not just what you take.

> *It's how you live, love, remember, and return to your truth.*

Integration Is the Real Medicine

To heal is to integrate.
To integrate is to become whole.
To become whole is to finally feel… **free**.

Let your healing not be an escape from the past, but a **homecoming to yourself**.
Let your embodiment be an act of **revolutionary softness**.
Let your integration be your **initiation into the new you**—the divine human you always were.

Activation Reflection: Integration as Sacred Completion

> **Sit in stillness.**
> **Breathe into the space behind your heart.**
> **Visualize every chapter of your journey—the wounds, the wisdom, the moments of expansion—becoming threads in a golden tapestry.**
>
> **Place your hands over your heart. Whisper aloud or within:**
> *I allow my healing to settle.*
> *I integrate all parts of myself.*
> *My light. My shadow. My love. My fear.*
> *I am no longer fragmented.*
> *I am whole. I am home.*
> *I am ready for what comes next.*
> And so it is.

CHAPTER 11: EXPLORING PARAPHYSICS AND THE ENERGETIC BLUEPRINT OF REALITY

"To understand the unseen is to understand the essence of reality itself."

What Is Paraphysics?

Paraphysics is the study of **subtle energy forces** and dimensions that **transcend the limitations of traditional physics**. It explores the energetic structure of the universe—beyond matter, beyond Newtonian mechanics—and into the **vibrational matrix** that underlies all form.

Unlike metaphysics, which philosophically explores reality beyond the physical, **paraphysics attempts to quantify** and understand the energy-based systems that influence and shape our lives.

Key focuses include:

- The nature of **energy fields** that interpenetrate and surround all living things
- **Interdimensional interactions** (how consciousness exists across multiple planes)
- The behavior of **non-local phenomena** (actions occurring outside of time and space)
- The science of **intentional energy manipulation** through thought, emotion, and frequency

Paraphysics bridges science and the sacred, offering

us the language to explain what mystics have practiced for centuries.

The Energetic Blueprint of the Human Body

Modern science acknowledges that the human body is not a closed, static system—it is an **open bioenergetic field**. According to biophysics, we are constantly interacting with electromagnetic fields and emitting measurable frequencies from the brain, heart, and nervous system.

The idea of an **energetic blueprint** suggests that before a cell forms or a condition manifests, there is already a **vibrational pattern** in place—like the architectural drawing before the structure is built.

This blueprint, often referred to as:

- The **morphogenetic field** (Rupert Sheldrake's term)
- The **light body** or **auric field**
- The **etheric double** in ancient teachings

...holds information about your health, emotions, and even your soul's intentions.

Scientific Insight:

- **HeartMath Institute** has shown that the heart emits an electromagnetic field that can influence others' emotional states.
- **Fritz-Albert Popp** discovered that cells emit **biophotons** —light particles that may carry bio-information across the body.

Spiritual Perspective:

Healers across cultures—from Indigenous shamans to Tibetan monks—speak of imbalances first appearing in the **energy body** before manifesting physically. Illness, therefore, can be seen as a **break in vibrational coherence**, rather than just a biological malfunction.

THE SUBTLE BODIES AND THEIR ROLES

The human energy system is composed of **multiple interpenetrating layers**, each corresponding to different frequencies of consciousness. These are often described in Vedic, Hermetic, and Kabbalistic traditions, and have now entered energy medicine lexicons worldwide.

Etheric Body:

- Closest to the physical
- Blueprint for cells and organs
- Holds **habitual patterns** and chronic conditions
- Trauma often registers here first as tension or fatigue

Emotional Body:

- Carries **emotional history**, especially suppressed or unresolved feelings
- Highly fluid, reactive to environment and relationships
- Imbalances here often manifest as anxiety, mood swings, or energetic overwhelm

Mental Body:

- Governs thought loops, belief systems, self-perception
- Holds the **mental programs** and conditioned patterns of

the ego

- When out of balance, can lead to obsessive thinking or mental fatigue

Astral Body:

- Activated during dreams, astral projection, or deep meditation
- Allows access to parallel timelines and the spirit realm
- Serves as a bridge between the personal self and the collective unconscious

Spiritual Body:

- Pure divine essence
- Contains soul purpose, karmic records, and higher wisdom
- Accessed during moments of transcendence, channeling, or deep inner stillness

Healing happens when we address **all levels**, not just the one we can touch.

Scalar Energy and Torsion Fields

Scalar Energy

Originally theorized by **James Clerk Maxwell** and later developed by **Nikola Tesla**, scalar energy is a **non-linear, non-Hertzian** form of energy that exists in the **zero-point field**—the space of stillness, potential, and infinite creation.

Unlike electromagnetic waves (which lose strength over distance), scalar waves are **non-local** and **can exist indefinitely** in coherent form.

Spiritual healers unknowingly work with scalar energy when they:

- Use **hands-on energy healing**
- Set strong intentions in Reiki or quantum touch
- Channel frequencies across time and space

Scalar energy is considered the **carrier wave** of consciousness, which is why prayers, affirmations, and visualizations—when emotionally charged—**create real results**.

Torsion Fields

Introduced by Russian physicist Nikolai Kozyrev, torsion fields are **spin-based energy fields** generated by movement, thought, and consciousness. They're believed to be the **mechanism behind psi phenomena**, including telepathy, precognition, and even intention-based healing.

> These energies explain why healing, manifestation, and intuition work—even when separated by time, space, or logic.

Practical Applications in Healing

Energy Medicine:

- Balances and realigns the biofield
- Often used alongside somatic or herbal healing

- Measurable shifts in HRV, stress markers, and inflammation have been observed

Vibrational Therapies:

- Utilize **sound and frequency** to re-pattern stuck energy
- Tibetan bowls, 432Hz music, solfeggio frequencies—each tuned to **restore coherence**
- Ideal for trauma release without verbal processing

Light & Color Therapy:

- Used in ancient Egypt, now rediscovered through chromotherapy and photobiomodulation
- Red light for mitochondrial repair, blue light for calming the mind, violet for spiritual clarity

Quantum Tools:

- Instruments like **Rife machines, bioresonance devices**, and **scalar wands** are being used in integrative clinics to scan and shift energy fields

Living as an Energetic Being

Once you understand your nature as an energetic being, **self-care evolves**:

You begin to track:

- Your **vibrational state**, not just your mood
- Your **auric hygiene**, not just your skin health

- Your **field boundaries**, not just your physical ones

You nourish your aura with:

- Sunlight, grounding, breath, and laughter
- Loving relationships and soulful environments
- Daily practices like smudging, energy brushing, or auric toning

You make decisions not just by logic, but by **energetic alignment**—asking, *"Does this expand or constrict my field?"*

> You no longer chase clarity—you tune to it.
> You don't seek protection—you *vibrate above interference.*

CONCLUSION

Closing the Loop – A Return to Self and Source

"The path was never about finding something outside of you. It was about remembering what you are—what you've always been."

Integration Is Return

We have journeyed through realms of quantum physics and cosmic intelligence. We have explored the psychic senses, ancestral lineages, and subtle bodies. We have remembered how timelines shift and how the soul expands. And now, we arrive here—not at an ending, but a spiral of sacred completion. A return. Not to a former version of ourselves, but to our **true Self**.

Integration is not a final step; it is a frequency of wholeness. It is what happens when the fragments of who we are, have been, and are becoming collapse into alignment. Integration means that all parts of you—the seeker, the wounded one, the mystic, the child, the warrior, the lover—stand together in harmony.

This chapter is an invitation to breathe, to absorb, to let the knowledge transform into embodied wisdom. You are not here to become something new. You are here to *remember*.

The Illusion of Separation

So much of our suffering comes from the illusion that we are separate: from each other, from Source, from our own divinity. But modern physics and ancient mysticism both agree: **separation is an illusion.**

Quantum entanglement shows us that particles, once connected, remain linked across space and time. Spiritual teachings tell us the same. We were never truly divided. We were only made to forget.

We were trained to see spirit as something distant. God as something above us. Healing as something outside of us. But when we strip away the layers of fear, shame, conditioning, and ego, we remember what has always been within:

> You are not just a creation of the Divine. You are the Divine made manifest.

This is the return. This is the homecoming.

Returning to Self

Returning to the Self is not about going back to who you were —it's about stepping fully into who you've always been beneath the noise, beneath the masks, beneath the programming.

To return to Self is to:

- Trust the quiet knowing within
- Choose authenticity over performance
- Welcome all emotions as sacred messengers
- Feel your intuition not as fantasy, but as divine intelligence
- Create from your essence, not your ego

Your Self—with a capital S—is the part of you that remembers you are more than your trauma, more than your titles, more than your roles. It is the eternal witness, the soul's compass.

Returning to Source

Source is not a place. It's not a person. It's not a religion or a doctrine. **It is the origin of all things and the essence of who you are.**

To return to Source is to:

- Recognize that everything is energy and vibration

- See love as the fundamental frequency of the universe
- Operate from compassion rather than judgment
- Witness the divine in others—and yourself
- Allow miracles to be part of your daily language

In this remembrance, Source becomes a lived reality. Not something to worship at a distance, but something to *embody*.

You are the frequency of Source encoded in human form.

The Sacred Practice of Closing the Loop

To close the loop is to consciously bring your healing journey into a full circle of completion. It doesn't mean the work ends. It means the struggle does. It means you move forward in grace, wisdom, and deep inner peace.

Here are practices to support this sacred closing:

1. Soul Journaling: Write a letter to the version of you who began this journey. Thank them. Forgive them. Celebrate them.

2. Ritual of Integration: Light a candle. Play music that opens your heart. Speak your truths aloud. Let your body know it is safe to live fully.

3. Source Invocation: Every morning, place your hands over your heart and say: "I return to myself. I return to Source. I am whole."

4. Embodied Action: Choose one daily action that reflects the wisdom you've gained—be it a boundary, a creation, or a sacred pause.

You Are the Portal

The end of this journey is not a closing door. It is a portal. A vortex of potential that opens when you claim yourself fully.

You are the one your ancestors dreamed of. You

> are the manifestation of prayers, whispers, and star-coded blueprints. You are the return.

So breathe deeply. Stand tall. Let the energy of this journey ripple through your cells.

You are not separate. You are not waiting. You are not unfinished.

You are the loop. You are the light. You are the living remembrance of the Divine.

And so it is.

EPILOGUE:

The Spiral Continues

"You were never meant to stay where you started. You were meant to evolve into the remembrance of who you truly are."

You've reached the final page, but not the end.

Because there is no end. There is only evolution. There is only expansion. There is only the eternal spiral of awakening and return.

What you've just read is not a book. It is an **invitation**.

An invitation to explore your mind not just as a brain, but as a multidimensional receiver of consciousness. An invitation to walk in your body not as a vessel of pain or performance, but as a sacred technology of remembrance. An invitation to remember your soul not as an abstract idea, but as a living field of energy encoded with ancient wisdom.

You are not separate from the universe—you *are* the universe. You are not a passive observer of reality—you *are* reality, shaping itself through your thoughts, your choices, your frequency.

This is not the end of your story. This is a **new beginning** at a higher frequency.

You now hold the keys. To healing. To consciousness. To Source.

Your frequency will continue to shift. Your timeline will continue to unfold. Your soul will continue to call you deeper and higher and freer.

And every time you feel lost, unworthy, fragmented, or confused, may you remember this truth:

> *You are not broken—you are becoming. You are not behind—you are awakening. You are not too late—you are right on divine time.*

As you walk forward from this page, may you embody what you've remembered:

- That your thoughts are waves of creation.
- That your feelings are messengers of frequency.
- That your being is both human and divine.

This is not the final chapter. This is a turning point. A rebirth. A recalibration. A reclamation.

So go. Create. Speak. Rest. Love. Heal. Guide. Shine.

And when you forget who you are, come back to this page. Come back to this remembrance:

> *You are Source made form. You are the field and the force. You are the light and the witness. You are here. Now. For a reason.*

Let this spiral continue. Let it lift you. Let it root you. Let it bring you home.

And so it is.

With love, Dr. Deilen Michelle Villegas

Appendix A: Glossary of Terms and Concepts

A

- **Akashic Records**: An etheric archive believed to contain all human experience, soul history, and cosmic memory.
- **Astral Body**: The energetic layer associated with dreams, spiritual travel, and emotional experiences.

B

- **Biofield**: The electromagnetic energy field that surrounds and interpenetrates the human body, often associated with the aura.
- **Biophotons**: Ultralow light emissions from cells that may communicate biological information.

C

- **Claircognizance**: Clear knowing. Intuitive knowledge that arises spontaneously.
- **Clairsentience**: Clear feeling. The ability to sense energies or emotions.
- **Clairvoyance**: Clear seeing. Intuitive perception of visuals beyond the physical realm.
- **Consciousness**: The state of being aware, often described in this text as a field or fundamental force.

D

- **Downloads**: Sudden influxes of insight, vision, or truth received energetically.

E

- **Etheric Body**: The energetic duplicate of the physical body, believed to hold patterns of health or disease.
- **Epigenetics**: The study of how environment and behavior influence gene expression.

F

- **Frequency**: The rate at which energy vibrates; linked to emotional states, consciousness, and healing.

M

- **Morphogenetic Field**: Hypothetical field containing the blueprint of biological forms and behavior.
- **Multidimensionality**: The concept that humans exist across multiple planes of reality simultaneously.

P

- **Paraphysics**: Study of energetic or metaphysical phenomena that extend beyond classical physics.
- **Parapsychology**: Scientific study of psychic abilities and other extrasensory phenomena.

Q

- **Quantum Field**: The energetic backdrop of the universe; a field of potential and probability.
- **Quantum Leaping**: A shift in one's state of being or timeline, often instant and profound.

S

- **Scalar Energy**: Non-linear, non-Hertzian energy theorized by Nikola Tesla; often linked to zero-point energy.
- **Soul Retrieval**: A shamanic or spiritual healing practice that reclaims fragmented parts of the self.
- **Subtle Body**: Layers of energetic anatomy (etheric, emotional, mental, etc.) surrounding the physical form.

Appendix B: Annotated Bibliography of Peer-Reviewed Studies

Chapter 1: The Science of Oneness

Aspect, A., Dalibard, J., & Roger, G. (1982). Experimental Test of Bell's Inequalities Using Time-Varying Analyzers. *Physical Review Letters, 49*(25), 1804–1807.

> This experiment confirmed quantum entanglement, illustrating how particles remain connected across space and time, supporting the concept of a unified quantum field.

McCraty, R., Atkinson, M., & Tomasino, D. (2001). Science of the Heart: Exploring the Role of the Heart in Human Performance. *HeartMath Institute.*

> Demonstrates heart-brain coherence and the bioelectromagnetic communication of the human heart, reinforcing the idea of energetic interconnectedness.

Chapter 3: The Brain as a Receiver

Hameroff, S., & Penrose, R. (2014). Consciousness in the universe: A review of the 'Orch OR' theory. *Physics of Life Reviews, 11*(1), 39–78.

> Offers a quantum theory of consciousness involving microtubules within neurons, supporting the idea that the brain may act as a non-local receiver of consciousness.

Dispenza, J. (2017). *Becoming Supernatural: How Common People Are Doing the Uncommon.* Hay House. *(Referenced data drawn from brain scan studies documented in the text.)*

> Contains case studies of neuroplastic shifts observed via EEG and HRV, supporting the ability of consciousness to rewire the brain.

Chapter 5: Extrasensory Perception & the Clairs

Targ, R., & Puthoff, H. E. (1974). Information Transmission Under Conditions of Sensory Shielding. *Nature, 252*, 602–607.

Foundational remote viewing experiment conducted at SRI, which demonstrated statistically significant extrasensory perception (ESP) under blinded conditions.

Bem, D. J. (2011). Feeling the Future: Experimental Evidence for Anomalous Retroactive Influences on Cognition and Affect. *Journal of Personality and Social Psychology, 100*(3), 407–425.

Explores precognition and retrocausality, suggesting that the human mind can access future events.

Chapter 6: Dreams and Divine Downloads

LaBerge, S. (1985). *Lucid Dreaming.* Ballantine Books.

Presents extensive research on lucid dreaming and its therapeutic potential, validating dreams as conscious states with transformative power.

Kuhn, G., & Brass, M. (2009). The Cognitive Neuroscience of Magic. *Trends in Cognitive Sciences, 13*(11), 505–509.

Discusses perception and expectation, indirectly supporting the flexibility of consciousness in altered dream-like states.

Chapter 7: Multidimensionality and Timeline Theory

Everett, H. (1957). "Relative State" Formulation of Quantum Mechanics. *Reviews of Modern Physics, 29*(3), 454–462.

The foundational paper proposing the Many-Worlds Interpretation, aligning with the concept of multiple coexisting timelines.

Yehuda, R. et al. (2015). Holocaust Exposure Induced Intergenerational Effects on FKBP5 Methylation. *Biological Psychiatry, 80*(5), 372–380.

Demonstrates that trauma is passed down epigenetically, supporting the idea that we are multidimensional beings influenced by timelines beyond our conscious awareness.

Chapter 10: Paraphysics and Scalar Energy

Popp, F.-A., Li, K. H., & Gu, Q. (1992). Recent Advances in Biophoton Research and Its Applications. *World Scientific Publishing.*

Presents biophoton theory as a medium of cellular communication, forming the basis for light-based healing.

Korotkov, K. G. (2002). Human Energy Field: Study with GDV Bioelectrography. *Journal of Alternative and Complementary Medicine, 8*(1), 111–117.

Details early gas discharge visualization technology used to measure the human aura, supporting the energetic anatomy model.

Kozyrev, N. A. (1991). Causal or Torsion Fields in a Physical Vacuum. *Russian Academy of Sciences.*

Describes torsion field theories used in paraphysical science to explain psychic phenomena, intuitive knowing, and remote healing.

Appendix C: Scientific and Reseach References by Chapter

Chapter 1: The Science of Oneness

- Bell's Theorem (John Stewart Bell, 1964)
- Quantum Entanglement (Aspect et al., 1981)
- HeartMath Institute Research on Heart-Brain Coherence

Chapter 2: The Collective Field

- Rupert Sheldrake's Morphic Resonance
- EEG studies on meditating groups (Davidson, Kabat-Zinn, et al.)

Chapter 3: The Brain as a Receiver

- Penrose & Hameroff's Orch-OR Theory of Consciousness
- Dr. Joe Dispenza's brain scans on meditation and neuroplasticity

Chapter 4: Intuition & Insight

- Studies on intuitive decision-making in military and corporate strategy (U.S. Naval Research)

Chapter 5: The 'Clairs' and ESP

- Rhine Research Center studies on extrasensory perception
- CIA Project Stargate declassified reports

Chapter 6: Dreams, Downloads & Divine Guidance

- Lucid dreaming research (Stephen LaBerge)
- Studies on hypnagogic and theta brain states

Chapter 7: Multidimensionality & Timelines

- Many-Worlds Interpretation (Hugh Everett)
- Epigenetic memory and inherited trauma (Rachel

Yehuda, 2015)

Chapter 8: Embodiment of the Divine Human

- Somatic Experiencing (Peter Levine)
- Polyvagal Theory (Stephen Porges)

Chapter 9: Integration and Quantum Healing

- Biophoton studies (Fritz-Albert Popp)
- Energy medicine and HRV regulation studies (HeartMath)

Chapter 10: Paraphysics and the Energy Body

- Nikola Tesla's scalar wave theories
- Kozyrev Mirrors and torsion fields
- Gas Discharge Visualization (Korotkov)

Chapter 11: Return to Source

- Cross-religious studies on mystical experience (William James)
- Near-Death Experience research (Dr. Eben Alexander)

Appendix D: Recommended Reading & Further Exploration

Books

- *The Field* by Lynne McTaggart
- *Becoming Supernatural* by Dr. Joe Dispenza
- *The Biology of Belief* by Dr. Bruce Lipton
- *The Holographic Universe* by Michael Talbot
- *Hands of Light* by Barbara Brennan
- *Waking the Tiger* by Peter Levine
- *The Power of Now* by Eckhart Tolle

Documentaries & Films

- *What the Bleep Do We Know!?* (2004)
- *Heal* (2017)
- *Superhuman: The Invisible Made Visible* (2020)

Courses & Institutions

- Institute of Noetic Sciences (IONS)
- HeartMath Institute
- The Monroe Institute
- Gaia.com educational series on metaphysics, energy healing, and consciousness

Appendix E: Practitioner Resources and Toolkits

Practitioner Toolkits

1. Energy Field Assessment Worksheet

- Identify energetic imbalances using language your clients understand.
- Sections include: physical sensations, emotional patterns, intuitive nudges, recurring dreams.
- Clients rate from 1–10 how strongly they feel each, offering a guide for energetic tuning.

2. Timeline Mapping Exercise

- A guided worksheet to help clients identify key turning points, karmic loops, and energetic patterns in their current and parallel lifelines.
- Supports deep reflection and visualization for healing and integration.

3. Soul Retrieval and Fragment Integration Prompts

- Journal prompts and meditative reflection questions to support parts work, trauma healing, and inner child integration.
- Sample prompt: "What part of you still feels unheard? What does it need to return home?"

4. Client Intake Form for Quantum Healing Practitioners

- Includes energetic history, spiritual background, primary emotional themes, and intuitive strengths.
- Encourages practitioners to track themes across sessions based on quantum frequency shifts.

5. Integration Aftercare Guide

- A take-home sheet for clients post-session with grounding tools, journaling suggestions, hydration and sleep reminders, and affirmations for energetic balance.

Recommended Practitioner Readings

- *Energy Medicine* by Donna Eden
- *The Subtle Body* by Cyndi Dale
- *The Body Keeps the Score* by Bessel van der Kolk
- *Quantum Healing* by Deepak Chopra
- *Anatomy of the Spirit* by Caroline Myss

Training & Certification Resources

For Trauma-Informed Practice

- Somatic Experiencing Institute (Peter Levine)
- The Polyvagal Institute (Stephen Porges)
- Internal Family Systems (IFS Institute)

For Quantum & Energy Healing

- Institute of Noetic Sciences (IONS)
- The Quantum University
- Biofield Tuning Academy
- The Monroe Institute

For Spiritual & Intuitive Development

- Intuitive Way (Penney Peirce)
- Gaia.com Consciousness Courses
- Shamanic Energy Medicine School (Alberto Villoldo)

Practitioner Self-Care Checklist

1. Daily grounding ritual (breathwork, nature, sound)
2. Boundaries and energetic hygiene practices
3. Supervision or peer support sessions
4. Regular body-based practices (yoga, dance, somatic movement)
5. Continuous education (reading, courses, mentorship)

You are not just a practitioner—you are a portal. Keep your channel clear, your heart aligned, and your purpose rooted in love.

Made in the USA
Monee, IL
05 November 2026

67882756R10089